Chemistry

A Concise Revision Course for CXC

ANNE TINDALE

STANLEY THORNES (PUBLISHERS) LTD

First published in 1992.
Second edition published in 1998 by:
Stanley Thornes (Publishers) Ltd
Ellenborough House
Wellington Street
CHELTENHAM GL50 1YW
England

A catalogue record of this book is available from the British Library.

ISBN 0-7487-3673-5

Typeset by Mathematical Composition Setters Ltd, Salisbury, Wiltshire.
Printed and bound in Great Britain by Redwood Books, Trowbridge, Wiltshire.

Contents

	Introduction	iv
	The key to success	v
	The Periodic Table	x
1	The states of matter	1
2	Elements, compounds and mixtures	5
3	Atomic structure	12
4	The Periodic Table	16
5	Chemical formulae and bonding	20
6	The structure of solids	28
7	Chemical equations	31
8	Solutions, solubility, suspensions and colloids	34
9	Acids, bases and salts	37
10	The mole and chemical calculations	46
11	Quantitative analysis calculations	56
12	Types of chemical reactions	60
13	Oxidation and reduction	62
14	Electrochemistry and electrolysis	68
15	Energy and chemical energetics	78
16	Rates of reaction	85
17	Metals and the reactivity series	89
18	The extraction and uses of metals	94
19	The chemistry and uses of some non-metals	100
20	Non-metals in living systems and the environment	107
21	Laboratory preparation and identification of gases	112
22	Some industrial processes	115
23	Qualitative analysis	119
24	An introduction to organic chemistry	125
25	The hydrocarbons	129
26	Alcohols, alkanoic acids and esters	136
27	Polymers	144
	Revision questions	149
	Numerical answers to revision questions	158
	Index	159

Introduction

This book is intended primarily as a revision text for students studying for the CXC Chemistry Examination. The facts are presented concisely and in such a form that they can be easily revised; to this end, the material in each chapter is presented under a number of sub-titles, key words are shown in **bold type**, and tables and diagrams have been used wherever possible. Several chapters also contain worked examples, in which particular emphasis is placed on *the mole* – this being one area that many students find particularly difficult.

Advice is given in the following section, 'The key to success', on examination technique, organising revision and School-based Assessment, together with an explanation of the format of the CXC Chemistry Examination. This advice and information should prove invaluable to the student.

Revision questions, organised topic-by-topic, have been included to enable the student to assess his or her own knowledge and understanding of each topic.

Whilst the text has been written principally for students, teachers should also find it extremely useful in reducing the time spent giving notes. This should then enable the teacher to develop a more experimental approach to the subject, as required by the CXC Chemistry syllabus.

This book and the CXC Chemistry syllabus

Obtain a copy of the **CXC Chemistry syllabus** which provides information on the course content topic-by-topic, School-based Assessment and the form of the examination. You will find each chapter in this book covers a specific topic in the syllabus in a form which is easy to revise.

- **Chapters 1 to 16** cover topics in **Section A** (Principles of Chemistry),
- **Chapters 17 to 23** cover **Section B.2** (Inorganic Chemistry),
- **Chapters 24 to 27** cover **Section B.1** (Organic Chemistry).

No attempt has been made to cover **Section C** since this section covers topics selected for special study, and these topics change on a regular basis.

At the end of this book, on pp. 149–158, you will find some **revision questions**. Each question aims to test a specific topic, as indicated by the headings. By working through these, you should be able to assess your knowledge and understanding of each topic in turn. In the CXC Examination, however, questions may test a knowledge of more than one topic.

Organising your revision

1 **Obtain copies of past CXC Chemistry Examination papers.**

2 **Revise on your own** – preferably in a room on your own.

3 **Revise in silence** – no radio, television, etc.

4 **Revise sitting upright at a table** – not in an easy chair or lying on a bed.

5 **Revise the entire syllabus topic-by-topic** – do not pick out isolated topics only.

6 **Plan a revision time-table** for eight weeks before the examination. This must cover weekends, evenings when you don't have a lot of homework, the Easter Vacation (which is *not* a vacation for you) and study leave. Divide full days into **three sessions**, morning, afternoon and evening. These should be a minimum of three hours each. Aim to study during **two sessions only** – do something completely different during the third. During a session, revise for **short periods**, one hour maximum, and take **short rests** of 5–10 minutes between.

7 **Aim to learn one topic per revision session** – more if topics are short.

8 **Work through topics in order** – when **all** have been learnt once, return to the first topic. Aim to cover each topic **several times**.

9 **Read through the entire topic** to ensure you **understand** it before attempting to learn it.

10 There are several **methods** which you can use to **learn** your work:

- **Read a topic several times** and then try to write out the **main points** without the book.

- **Practise writing equations** wherever relevant. Do not attempt to learn all equations 'parrot fashion' – learn **how** to write them, e.g. if you learn that when a base reacts with an acid it forms a salt and water, then you can write an equation for the reaction between **any base** and **any acid**.

- **Use memory aids** such as **mnemonics**, e.g. the order of metals in the reactivity series: **p**eople **s**ometimes **c**ollect **m**agazines **a**bout **z**oos **i**n **L**ondon **h**aving **c**ute **m**onkeys and **s**mall **g**orillas; **p**otassium, **s**odium, **c**alcium, **m**agnesium, **a**luminium, **z**inc, **i**ron, **l**ead, (**h**ydrogen), **c**opper, **m**ercury, **s**ilver, **g**old,
or **word associations**, e.g. a**n**ions – **n**egative (cations must, therefore, be positive).

- **Test yourself** using relevant questions on pp. 149–158 and others from past CXC Examination papers.

- **Do not try to learn notes 'parrot fashion'** – notes learnt this way are rarely understood and most questions test **understanding** not repetition.

Personal preparation for the examination

THE NIGHT BEFORE THE EXAMINATION

- Prepare all necessary equipment: pen (plus a spare), sharp HB pencil, pencil sharpener, ruler, rubber, calculator.

- Go to bed early and get a good night's sleep.

- No last-minute revision.

THE MORNING OF THE EXAMINATION

- Get up early and refreshed.

- Eat a good breakfast.

- Leave home early so you arrive at the examination room in plenty of time.

- No last-minute revision.

The CXC Chemistry examination

The examination consists of **three papers** which evaluate your performance under the following headings:

- **Knowledge and comprehension**

- **Use of knowledge**

- **Experimental skills**.

PAPER 1 (1 $\frac{1}{4}$ HOURS)

This contains **60 multiple-choice questions** covering Sections A, B.1 and B.2. **Four choices** of answer are provided for each question, one is correct. If you don't know the correct answer, work it out by **eliminating** the incorrect answers. Read the questions thoroughly, as some may be in the **negative**:

Sample question

An atom of an element contains 12 neutrons and 11 electrons. Which of the following is *incorrect*?

A The relative atomic mass of the element is 23.
B An atom of the element contains 11 protons.
C The atomic number of the element is 12.
D The element is in Group I of the Periodic Table.

PAPER 2 (1 $\frac{1}{2}$ HOURS)

This paper consists of at least **five compulsory structured questions**, each divided into several parts. Usually one question tests topics from Section B.1 of the syllabus and the others test topics from Sections A and B.2. The questions usually begin with some information, e.g. a diagram of a piece of apparatus, a table, a graph or an equation, which must be analysed when answering. One question is always a **data analysis question**. As implied, this question provides you with a set of **data** to be analysed, such as the results obtained from a practical investigation.

Answers may be a word, a sentence or a paragraph, and these are to be written in **spaces** provided on the paper. The spaces indicate the length of answer required and answers **must** be restricted to them.

PAPER 3 (1 HOUR + 10 MINUTES READING TIME)

This contains **six guided essay-type questions**, each divided into several parts. The questions are in **pairs**; **one** question **from each pair** must be answered. The first four questions test topics from Sections A and B – usually two of these test topics from Section A, one tests topics from Section B.1 and one tests topics from Section B.2. The last pair of questions tests topics in Section C – be sure you have a thorough knowledge of the special topic you studied.

The questions on Paper 3 require a greater element of **essay writing** than those on Paper 2 and each question should take **20 minutes**.

Graphs

When drawing graphs:

- Choose a **scale** which is easy to work with and makes maximum use of the graph paper.

- Enter **numbers** along the axes and **label** the axes clearly, stating relevant units, e.g. °C, g.

- Use **small dots** to plot points.

- Draw the **best** straight line or smooth curve – it need not necessarily pass through all the points.
- Give the graph an accurate **title**.

Examination technique

1 **Read the instructions** at the beginning of each paper **carefully** and do **exactly** what they say.

2 **Be sure you know the number of questions to be answered** and attempt that number – do not answer more or fewer than required.

3 **Make maximum use of the 10 minutes reading time** on Paper 3 to choose the **correct questions** to answer – use the mark allocations to choose those which will give you the best marks.

4 **Read each question several times** before answering to be sure you **understand** exactly what it is asking.

5 **Underline key words in the question** – this will help you to answer exactly what the question asks.

6 **Give only the information required** – no marks are awarded for padded-out, irrelevant facts. The length of answer needed is often suggested by the marks awarded for it.

7 **Give answers which are precise and factual**, and use scientific terminology correctly.

8 **Chemical equations** must always be **balanced** and include **state symbols**.

9 **Show all working** when answering questions requiring **calculations** – remember that a numerical answer must never be given without the appropriate **unit**.

10 **Time yourself accurately** – work out the time available for each question and **stick rigidly to it**, e.g. three questions in 60 minutes is 20 minutes per question.

11 **Never leave the room before the end of the examination** – if you do, you have not done your best.

Terms used on examination papers

Account for: give reasons for the facts.

Calculate: give a numerical answer which shows all relevant working.

Compare: pick out similarities and differences.

Contrast: pick out differences.

Define: state the precise meaning only.

Describe: give a detailed account which includes all relevant points.

Determine: using the information given, find a solution, usually by calculation.

Discuss: give a balanced argument which includes reasons for and against.

Distinguish between: pick out differences.

Evaluate: determine the value of the point in question.

Explain: give a clear, detailed account of the facts, which includes the reasons behind them.

Illustrate: make the answer clear by means of examples or diagrams.

Justify: give adequate grounds for your reasoning.

Outline: give an account which includes only the main points.

State or list: give brief, precise facts.

Suggest: put forward an idea.

Tabulate: construct a table using data provided or obtained.

NB When asked to 'give your opinions or views on', 'comment on' or 'give your reactions to', your answer must be based on your **chemical knowledge** and you must consider **both sides** of an argument.

A *word about* *School-based* *Assessment* (SBA)

SBA assesses you in the **experimental skills** and the **analysis and interpretation skills** that are involved in laboratory and field work. The assessments are carried out by your teacher during the two-year programme. SBA is worth **20%** of your final examination score and you have every opportunity to score highly, since assessments are made during normal practical classes and not under examination conditions. You must, therefore, make a **consistent effort** throughout the two years and not just in the final 'run-up' to the examination.

The Periodic Table

Relative atomic mass —— 1

Atomic (proton) number —— 1 H hydrogen

	Group I	Group II														Group III	Group IV	Group V	Group VI	Group VII	Group 0
	Alkali metals	Alkaline earth metals																		Halogens	Noble gases
Period 1																					4 He 2 helium
Period 2	7 Li 3 lithium	9 Be 4 beryllium														11 B 5 boron	12 C 6 carbon	14 N 7 nitrogen	16 O 8 oxygen	19 F 9 fluorine	20 Ne 10 neon
Period 3	23 Na 11 sodium	24 Mg 12 magnesium														27 Al 13 aluminium	28 Si 14 silicon	31 P 15 phosphorus	32 S 16 sulphur	35.5 Cl 17 chlorine	40 Ar 18 argon
Period 4	39 K 19 potassium	40 Ca 20 calcium	45 Sc 21 scandium	48 Ti 22 titanium	51 V 23 vanadium	52 Cr 24 chromium	55 Mn 25 manganese	56 Fe 26 iron	59 Co 27 cobalt	59 Ni 28 nickel	64 Cu 29 copper	65 Zn 30 zinc	70 Ga 31 gallium	73 Ge 32 germanium	75 As 33 arsenic	79 Se 34 selenium	80 Br 35 bromine	84 Kr 36 krypton
Period 5	85 Rb 37 rubidium	88 Sr 38 strontium	89 Y 39 yttrium	91 Zr 40 zirconium	93 Nb 41 niobium	96 Mo 42 molybdenum	98 Tc 43 technetium	101 Ru 44 ruthenium	103 Rh 45 rhodium	106 Pd 46 palladium	108 Ag 47 silver	112 Cd 48 cadmium	115 In 49 indium	119 Sn 50 tin	122 Sb 51 antimony	128 Te 52 tellurium	127 I 53 iodine	131 Xe 54 xenon
Period 6	133 Cs 55 caesium	137 Ba 56 barium	139 La 57 lanthanum*	178.5 Hf 72 hafnium	181 Ta 73 tantalum	184 W 74 tungsten	186 Re 75 rhenium	190 Os 76 osmium	192 Ir 77 iridium	195 Pt 78 platinum	197 Au 79 gold	201 Hg 80 mercury	204 Tl 81 thallium	207 Pb 82 lead	209 Bi 83 bismuth	210 Po 84 polonium	210 At 85 astatine	222 Rn 86 radon
Period 7	223 Fr 87 francium	226 Ra 88 radium	227 Ac 89 actinium†															

*58–71 Lanthanum series

140 Ce 58 cerium	141 Pr 59 praseodymium	144 Nd 60 neodymium	147 Pm 61 promethium	150 Sm 62 samarium	152 Eu 63 europium	157 Gd 64 gadolinium	159 Tb 65 terbium	162 Dy 66 dysprosium	165 Ho 67 holmium	167 Er 68 erbium	169 Tm 69 thulium	173 Yb 70 ytterbium	175 Lu 71 lutetium

†90–103 Actinium series

232 Th 90 thorium	231 Pa 91 protactinium	238 U 92 uranium	237 Np 93 neptunium	242 Pu 94 plutonium	243 Am 95 americium	247 Cm 96 curium	249 Bk 97 berkelium	251 Cf 98 californium	254 Es 99 einsteinium	253 Fm 100 fermium	256 Md 101 mendelevium	254 No 102 nobelium	257 Lr 103 lawrencium

1 The states of matter

Matter is anything that has **mass** and occupies **space**. All matter is composed of **particles** (see p. 2) and can exist in **three** states:

- Solid
- Liquid
- Gas.

Property	Solid	Liquid	Gas
Volume	Definite	Definite	Variable – expands to fill container
Shape	Definite	Takes shape of container bottom, surface is always horizontal	Takes shape of entire container
Expansion/ compression	Very difficult to expand or compress	Can be expanded or compressed slightly	Very easily expanded or compressed
Arrangement of particles	Packed closely together in a regular way	Randomly arranged with small spaces between	Randomly arranged with large spaces between
Forces of attraction between particles	Strong	Fairly weak	Very weak
Movement of particles	Vibrate – possess very small amounts of kinetic energy	Move slowly – possess medium amounts of kinetic energy	Move rapidly – possess large amounts of kinetic energy
Two-dimensional representation of particles			

Table 1.1 The three states of matter compared

Changing state

A substance can exist in any of the three states depending on conditions of **temperature** and **pressure**.

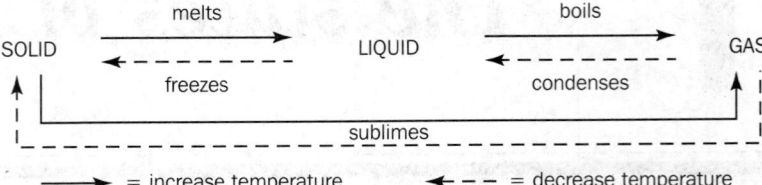

Fig. 1.1 Changing state

- If the temperature of a solid is measured as it is heated and changes state to a liquid and then from a liquid to a gas, a **heating curve** is obtained.
- If the temperature of a gas is measured as it is cooled and changes state to a liquid and then from a liquid to a solid, a **cooling curve** is obtained.

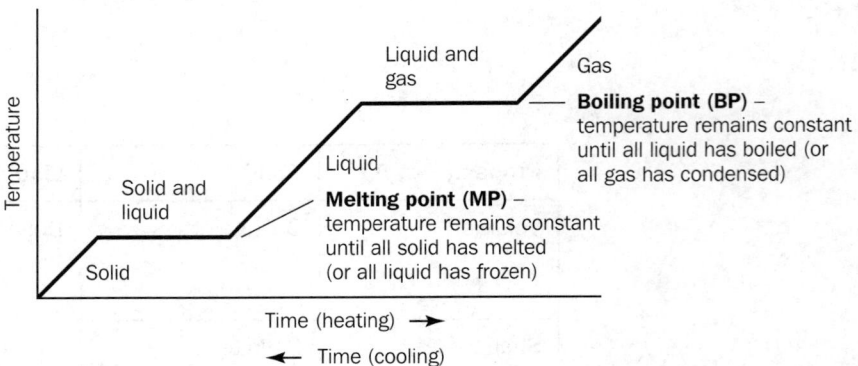

Fig. 1.2 A heating (or cooling) curve

- The **melting point** is the temperature at which a solid **melts** to form a liquid or a liquid **freezes** to form a solid.
- The **boiling point** is the temperature at which a liquid **boils** to form a gas or a gas **condenses** to form a liquid.

Matter is made of particles

The following provide **evidence** to support the theory that all matter is made of **particles**:

1 CRYSTALS HAVE A REGULAR SHAPE

All crystals of the same substance have **straight edges**, **flat surfaces** and the **same shape**, e.g. all sodium chloride crystals (common salt) are cubic.

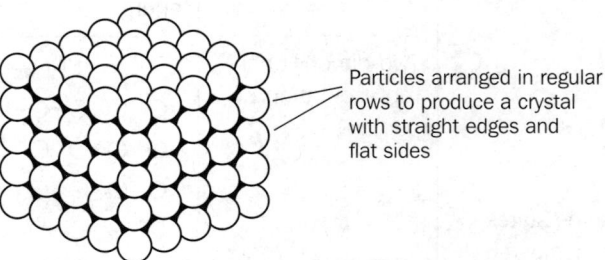

Particles arranged in regular rows to produce a crystal with straight edges and flat sides

Fig. 1.3 Why crystals have regular shapes

2 CRYSTALS DISSOLVE

When crystals are placed in water they **dissolve**, e.g. potassium manganate(VII) crystals dissolve in water to give a uniformly purple solution.

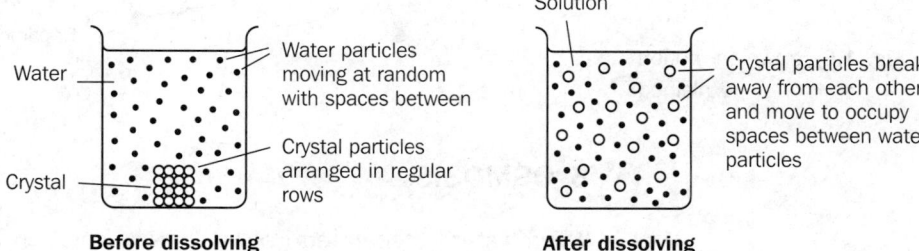

Fig. 1.4 Why crystals dissolve

Before dissolving

After dissolving

Diffusion is the movement of particles from a region of high concentration to a region of low concentration until they are evenly distributed.

3 GASES DIFFUSE

When two gases are put together they **mix** very quickly, e.g. when a jar of air is inverted over a jar containing brown bromine vapour, the brown vapour rises rapidly to fill the jar above.

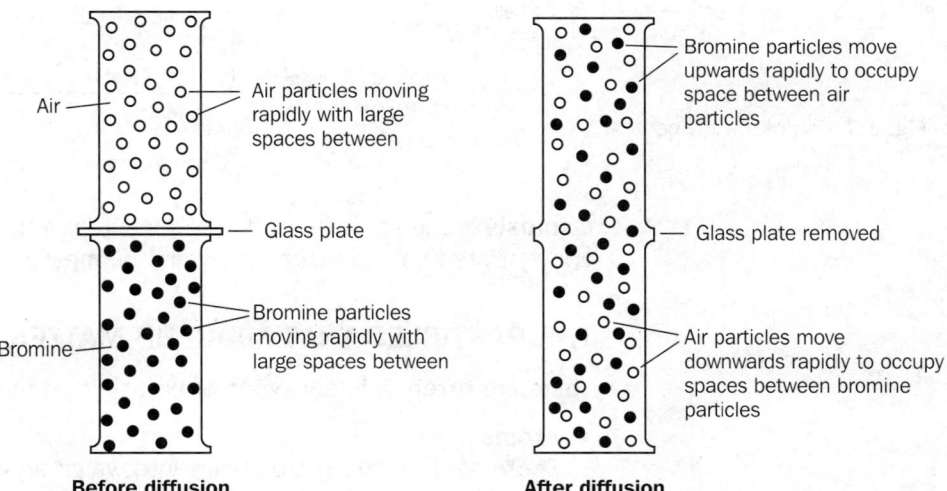

Fig. 1.5 Why gases diffuse

Before diffusion

After diffusion

4 BROWNIAN MOTION

When a beam of light is shone through dust or smoke, the specks of dust or smoke can be seen **moving** in an irregular way.

Brownian motion is named after Robert Brown who, in 1827, noted that pollen grains suspended in water moved in an irregular way as a result of being bumped by randomly moving water particles. (See Fig. 1.6 on next page.)

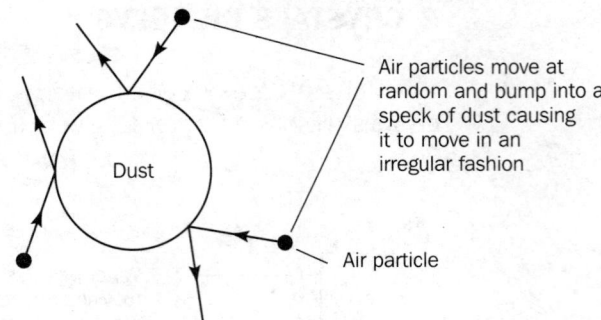

Fig. 1.6 Brownian motion
explained

5 OSMOSIS

When water is separated from a sucrose solution by a **semi-permeable membrane**, the water flows through the membrane into the sucrose solution, but the sucrose solution does not flow into the water. The volume of the sucrose solution **increases** and the volume of the water **decreases**.

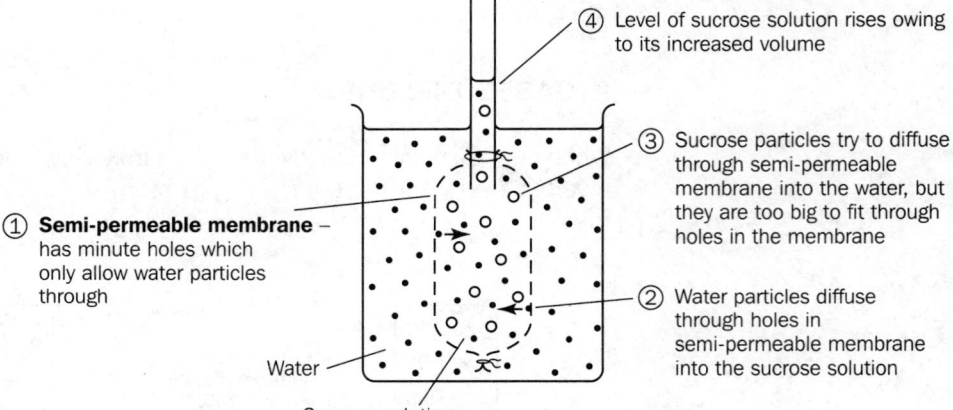

Fig. 1.7 Osmosis explained

Osmosis is the movement of a solvent from a more dilute to a more concentrated solution through a semi-permeable membrane.

THE PARTICLES THAT MAKE UP MATTER

There are **three** different types of particle that make up matter:

- **Atoms**
  These are the smallest particles into which an **element** can be divided (see p. 12).

- **Molecules**
  These are groups of atoms which are **bonded** together and which can exist as separate units. Atoms within a molecule may be of the same kind, e.g. oxygen O_2, or of different kinds, e.g. water, H_2O (see p. 24).

- **Ions**
  These are **electrically charged** particles formed from atoms, or groups of atoms bonded together, e.g. the sodium ion, Na^+; the sulphate ion, SO_4^{2-} (see p. 21).

2 Elements, compounds and mixtures

Elements

An **element** is a pure substance which cannot be split into simpler substances by any chemical process.

An element may consist of **individual atoms** of the same kind, e.g. copper, Cu, or **molecules** composed of atoms of the same kind, e.g. oxygen, O_2.

There are 103 different elements; they can be divided into **metals** and **non-metals**. Each element may be represented by an **atomic symbol** which represents **one atom** of the element.

Non-metals			
Element	**Atomic symbol**	**Element**	**Atomic symbol**
Argon	Ar	Nitrogen	N
Boron	B	Oxygen	O
Bromine	Br	Phosphorus	P
Carbon	C	Silicon	Si
Chlorine	Cl	Sulphur	S
Fluorine	F		
Helium	He		
Hydrogen	H		
Iodine	I		
Krypton	Kr		
Neon	Ne		
Metals			
Element	**Atomic symbol**	**Element**	**Atomic symbol**
Aluminium	Al	Magnesium	Mg
Barium	Ba	Manganese	Mn
Beryllium	Be	Mercury	Hg
Calcium	Ca	Nickel	Ni
Chromium	Cr	Potassium	K
Cobalt	Co	Silver	Ag
Copper	Cu	Sodium	Na
Gold	Au	Tin	Sn
Iron	Fe	Uranium	U
Lead	Pb	Zinc	Zn
Lithium	Li		

Table 2.1 The common elements and their atomic symbols

Property	Metals	Non-metals
State at room temperature	Solid (except mercury)	Solid, liquid or gas
Appearance of solid	Shiny	Dull
'Bendability' of solid	Malleable (can be hammered) and ductile (can be drawn out)	Brittle
Density	Generally high	Generally low
Melting and boiling points	Generally high	Generally low
Electrical and thermal conductivity	Good	Poor (except graphite, a form of carbon)
Ions formed	Cations (positive)	Anions (negative)
Oxidising or reducing agent	Reducing agent	Oxidising agent
Nature of oxide	Basic or amphoteric (see p. 104)	Acidic or neutral

Table 2.2 A comparison of the properties of metals and non-metals

Compounds

A *compound is a substance containing two or more different types of element bonded together chemically in fixed proportions and in such a way that their properties are changed.*

For example: magnesium + oxygen ⟶ magnesium oxide
            (element)   (element)   (compound)

The properties of magnesium oxide are different from those of magnesium and oxygen.

Compounds may be represented by **chemical formulae** (see p. 20).

Mixtures

A *mixture consists of two or more substances (elements and/or compounds) combined together in varying proportions. Each component retains its own independent properties and has undergone no chemical reaction with any other substance in the mixture.*

The components of mixtures may be **separated** using a variety of techniques, the technique used depends on the **properties** of the components (see pp. 7—10)

Pure substances

A **pure substance** consists of only **one** type of material. Pure substances have **fixed** melting and boiling points. Impurities **lower** the melting point and **raise** the boiling point of a pure substance. To find out if a substance is pure, its melting or boiling point is determined.

Separating mixtures

FILTRATION

Filtration is used to separate a liquid from a suspended or settled solid, e.g. soil from water.

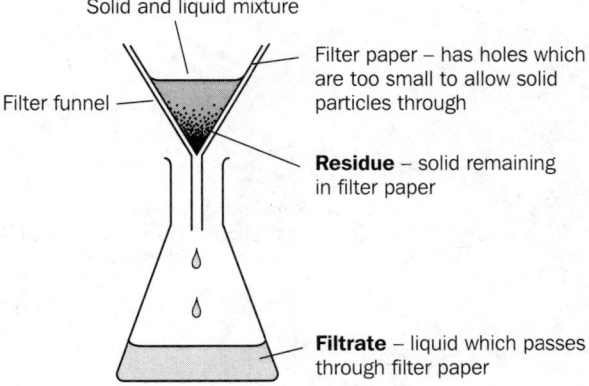

Fig. 2.1 Filtration

EVAPORATION

Evaporation is used to separate and retain the solid solute (see p. 34) from a solution if the solute does not decompose on heating or contain water of crystallisation, e.g. sodium chloride from sodium chloride solution.

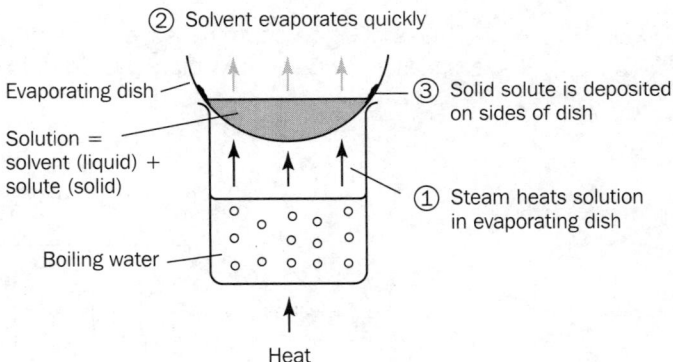

Fig. 2.2 Evaporation

CRYSTALLISATION

Crystallisation is used to separate and retain the solid solute from a solution, especially if the solid contains water of crystallisation, e.g. copper sulphate from copper sulphate solution.

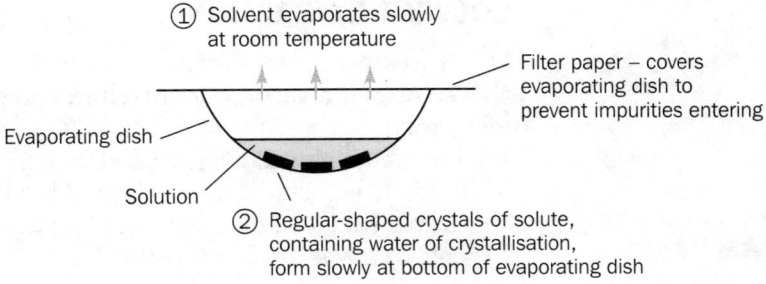

Fig. 2.3 Crystallisation

SEPARATING FUNNEL

A **separating funnel** is used to separate two **immiscible** liquids, e.g. oil and water. Immiscible liquids are liquids which do not mix.

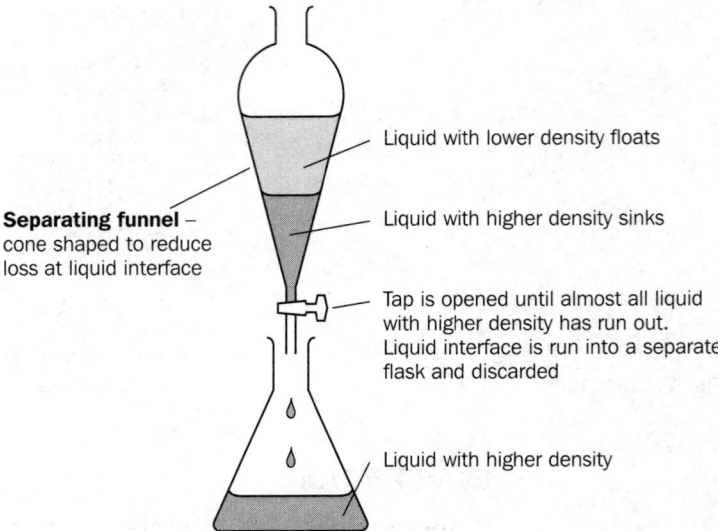

Separating funnel – cone shaped to reduce loss at liquid interface

Liquid with lower density floats

Liquid with higher density sinks

Tap is opened until almost all liquid with higher density has run out. Liquid interface is run into a separate flask and discarded

Liquid with higher density

Fig. 2.4 Separating funnel

PAPER CHROMATOGRAPHY

This is used to separate several solutes, usually coloured, present in a solution, e.g. dyes in black ink, pigments in chlorophyll.

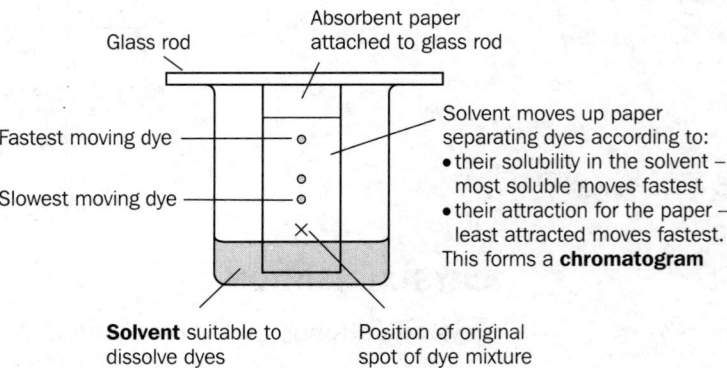

Glass rod

Absorbent paper attached to glass rod

Fastest moving dye

Slowest moving dye

Solvent moves up paper separating dyes according to:
• their solubility in the solvent – most soluble moves fastest
• their attraction for the paper – least attracted moves fastest.
This forms a **chromatogram**

Solvent suitable to dissolve dyes

Position of original spot of dye mixture

Fig. 2.5 Ascending paper chromatography

SOLVENT EXTRACTION

This is used to separate two solutes from a solution when one of the solutes is also soluble in a second, **immiscible solvent**. For example, water and trichloroethane are immiscible. If trichloroethane is added to an aqueous solution containing iodine and sodium chloride, the iodine dissolves in the trichloroethane, leaving the sodium chloride dissolved in the water. The two immiscible solutions can then be separated using a **separating funnel** and the solvents removed by **evaporation**.

SUBLIMATION

Sublimation is used to separate and retain a solid which sublimes from a mixture of solids, e.g. to separate iodine or ammonium chloride from a mixture containing either substance.

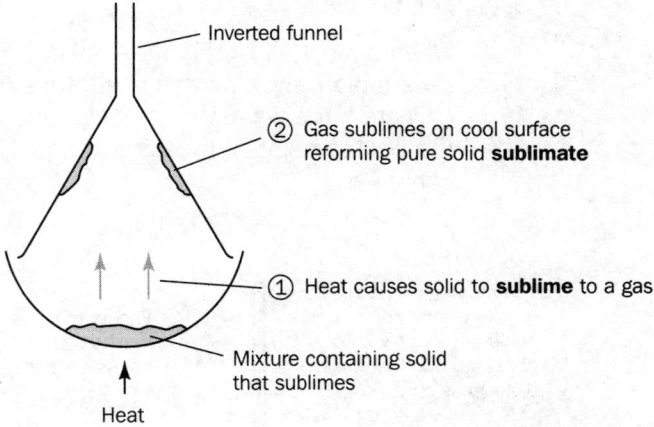

Fig. 2.6 Sublimation

SIMPLE DISTILLATION

Simple distillation is used to separate and retain the solvent from a solution, e.g. to obtain distilled water from tap or sea water. If impurities are not present, the solute may also be retained by **evaporation** or **crystallisation** of the solution remaining after distillation.

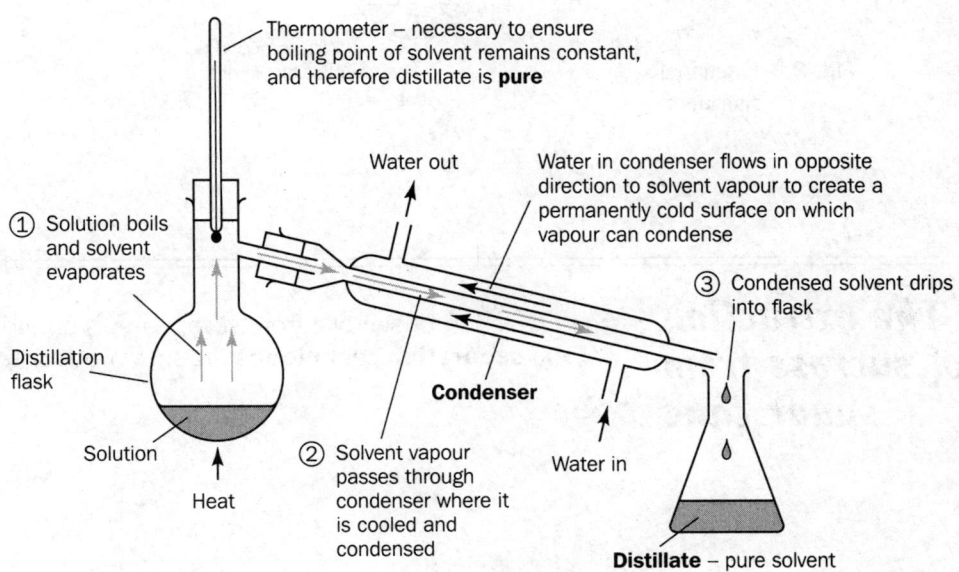

Fig. 2.7 Simple distillation

FRACTIONAL DISTILLATION

Fractional distillation is used to separate two **miscible** liquids which have **different** boiling points, e.g. separating ethanol, BP 78 °C, and water, BP 100 °C. Miscible liquids are liquids which mix completely.

As the mixture of vapours passes up the fractionating column it continually condenses and evaporates. This causes it to become increasingly richer in the **more volatile** component (the one with the lower boiling point), until the vapour reaching the top consists almost entirely of the more volatile component.

When almost all of the more volatile liquid has distilled over, the temperature rises rapidly showing that a **mixture** of both liquids is distilling over. This is collected in a separate container and discarded. Once the temperature reaches the boiling point of the second liquid, that liquid is then distilled into another container.

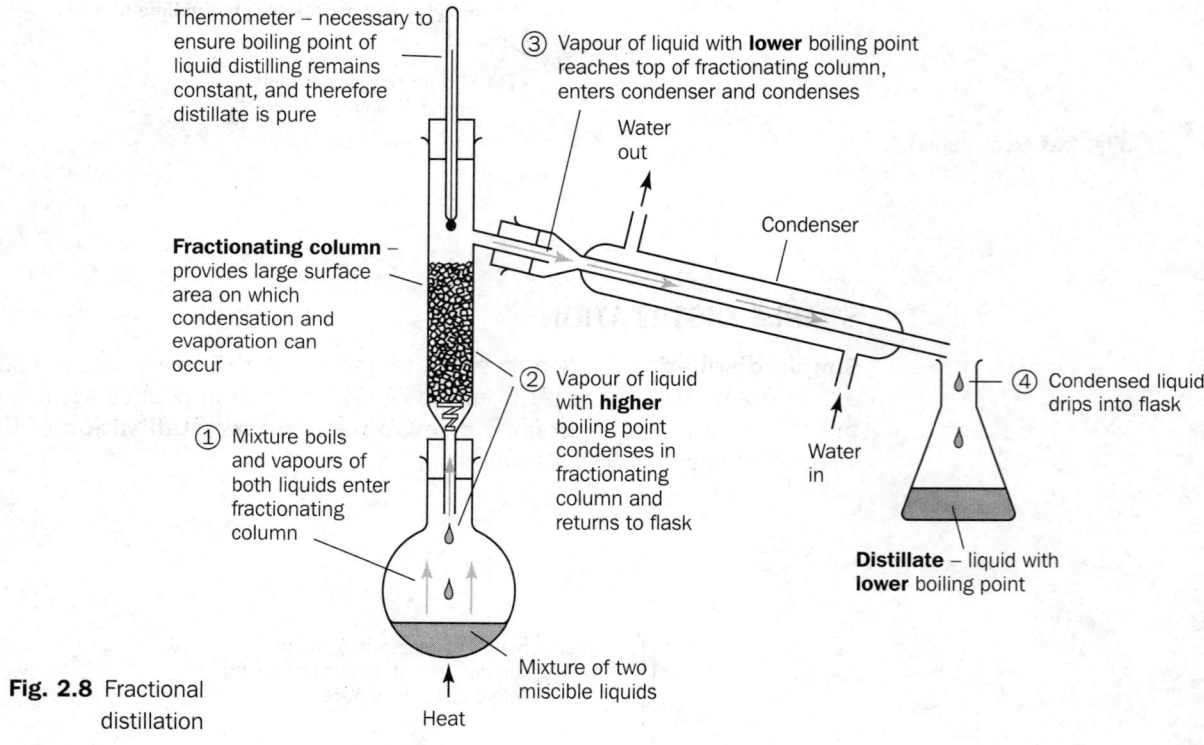

Thermometer – necessary to ensure boiling point of liquid distilling remains constant, and therefore distillate is pure

③ Vapour of liquid with **lower** boiling point reaches top of fractionating column, enters condenser and condenses

Water out

Condenser

Fractionating column – provides large surface area on which condensation and evaporation can occur

② Vapour of liquid with **higher** boiling point condenses in fractionating column and returns to flask

④ Condensed liquid drips into flask

① Mixture boils and vapours of both liquids enter fractionating column

Water in

Distillate – liquid with **lower** boiling point

Fig. 2.8 Fractional distillation

Heat

Mixture of two miscible liquids

The extraction of sucrose from sugar cane

The extraction of sucrose from sugar cane is an industrial process which utilises several **separation techniques**, as shown opposite.

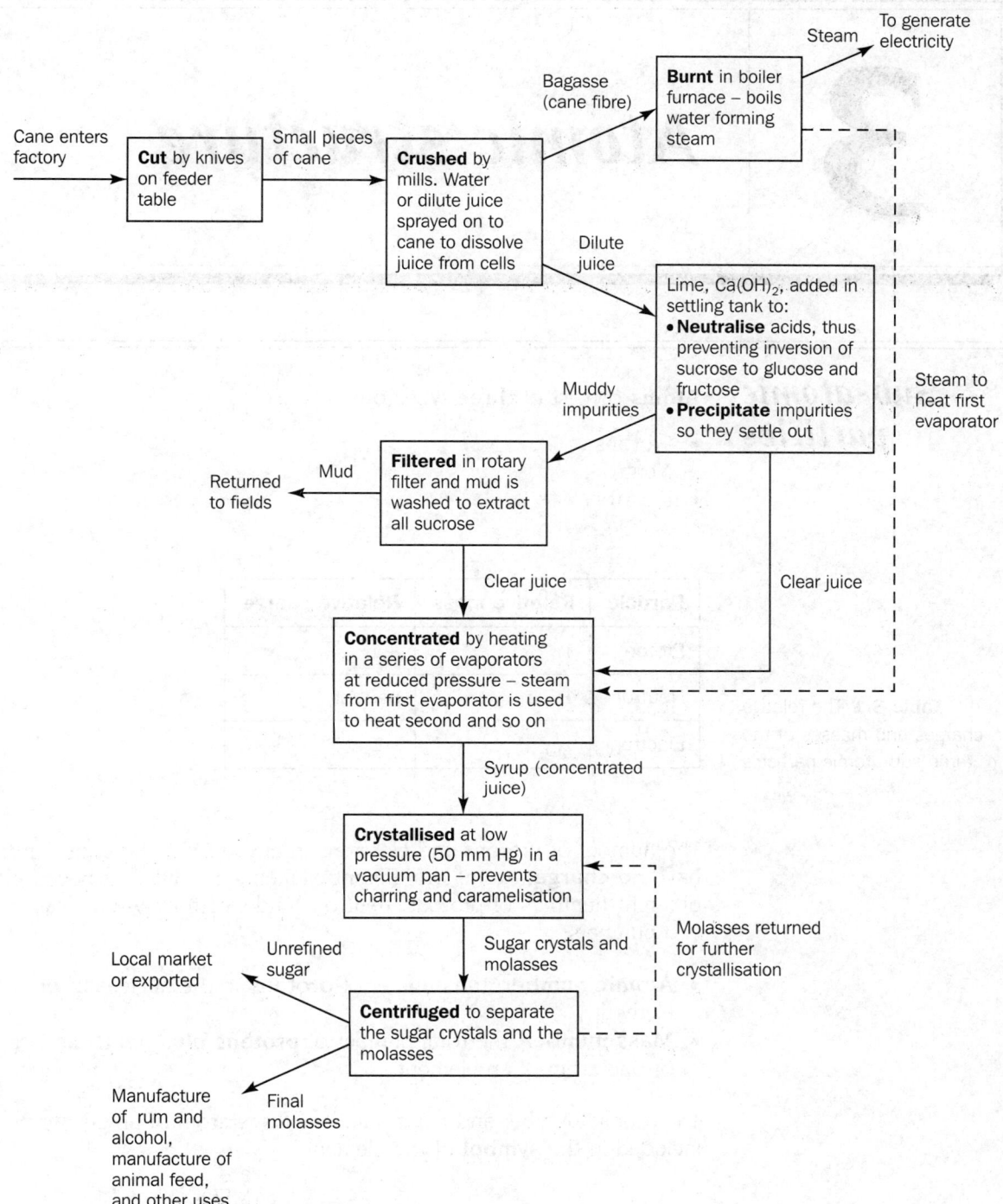

Fig. 2.9 A flow chart showing the extraction of sucrose from sugar cane

3 Atomic structure

Sub-atomic particles

Atoms consist of **three** types of particle:

- Protons
- Neutrons
- Electrons.

Particle	Relative mass	Relative charge
Proton	1	+1
Neutron	1	neutral
Electron	$\frac{1}{1840}$	−1

Table 3.1 The relative charges and masses of the three sub-atomic particles

The number of protons and electrons in any atom is the **same**, atoms therefore have **no charge**. Atoms of different elements are different because they contain different numbers of protons, neutrons and electrons. Atoms can be assigned two numbers:

- **Atomic number**: the number of **protons** in the nucleus of one atom of an element.
- **Mass number**: the total number of **protons plus neutrons** in the nucleus of one atom of an element.

The atomic number and mass number of one atom of an element can be included in the **symbol** of the element:

$$^{\text{mass number}}_{\text{atomic number}}\text{SYMBOL}$$

Example

$^{23}_{11}\text{Na}$ represents one atom of sodium:

Mass number = 23

Atomic number = 11

Number of protons = 11

Number of electrons = 11

Number of neutrons = 23 − 11 = 12

Arrangement of sub-atomic particles

Atoms are composed of **two** parts:

1 A central **nucleus** containing the **protons** and **neutrons** tightly packed together.

2 One or more **shells** or **energy levels** surrounding the nucleus containing **electrons** revolving at high speeds:
- Each shell is a **fixed distance** from the nucleus.
- Electrons in the shells closest to the nucleus have the **least energy**.
- Shells closest to the nucleus **fill up first**.
- Each shell contains up to a **fixed number** of electrons:
  The first shell (K shell) contains up to **2** electrons.
  The second shell (L shell) contains up to **8** electrons.
  The third shell (M shell) may be considered to contain up to **8** electrons at this level of study.

Further shells contain more electrons but are not studied here.

Examples

The hydrogen atom

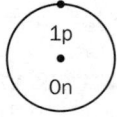

$^{1}_{1}$H contains 1 proton

           1 electron — arrangement 1

           0 neutrons

The fluorine atom

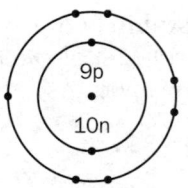

$^{19}_{9}$F contains 9 protons

           9 electrons — arrangement 2.7

           10 neutrons

The aluminium atom

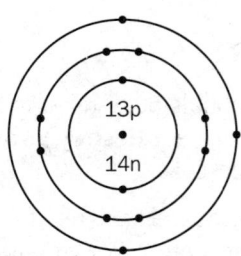

$^{27}_{13}$Al contains 13 protons

           13 electrons — arrangement 2.8.3

           14 neutrons

The **chemical properties** of elements are determined by the **number** and **arrangement** of electrons.

Isotopes

Isotopes are atoms of the same element with the same number of protons and electrons but different numbers of neutrons, i.e. they have the same atomic number but different mass numbers.

Examples

- **Chlorine** has **two** isotopes, $^{35}_{17}Cl$ and $^{37}_{17}Cl$. Naturally occurring chlorine consists of:

  75% $^{35}_{17}Cl$ — 17 protons, 17 electrons, 18 neutrons

  25% $^{37}_{17}Cl$ — 17 protons, 17 electrons, 20 neutrons

- **Hydrogen** has **three** isotopes, ^1_1H, ^2_1H and ^3_1H. Naturally occurring hydrogen consists of:

  99.985% ^1_1H, **protium** — 1 proton, 1 electron, 0 neutrons

  0.015% ^2_1H, **deuterium** — 1 proton, 1 electron, 1 neutron

The third isotope, **tritium**, ^3_1H, is man-made and radioactive. A tritium atom contains 1 proton, 1 electron, 2 neutrons.

- **Carbon** has **three** isotopes, $^{12}_6C$, $^{13}_6C$ and $^{14}_6C$. Naturally occurring carbon consists of:

  98.89% $^{12}_6C$, carbon-12 — 6 protons, 6 electrons, 6 neutrons

  1.10% $^{13}_6C$, carbon-13 — 6 protons, 6 electrons, 7 neutrons

  0.01% $^{14}_6C$, carbon-14: radioactive — 6 protons, 6 electrons, 8 neutrons

Since the **number** and **arrangement** of electrons in all isotopes of one element are identical, all isotopes of one element have the **same chemical properties**.

RADIOACTIVE ISOTOPES

The atoms of some isotopes are **unstable** and they split up spontaneously to form smaller atoms. As the nucleus splits, it also releases **radiation**. Isotopes which do this are called **radioisotopes** and they are said to be **radioactive**.

There are **three** types of **radiation**:

- α-**particles**: helium nuclei with a charge of $+2$.
- β-**particles**: electrons with a charge of -1.
- γ-**rays**: high energy electromagnetic radiation.

USES OF RADIOISOTOPES

1 γ-radiation from radioactive cobalt-60 is used in **cancer treatment (radiotherapy)**. Cancerous cells are destroyed by directing a controlled beam of this γ-radiation at the cells.

2 Very small quantities of radioisotopes are used as **tracers** in medical investigations and biological research. For example, radioactive iodine-131 is given to patients with defective thyroid glands; the path of the radioactive iodine can then be followed through the body. Radioactive carbon-14 is used to study photosynthesis in plants.

3 Radioactive plutonium-238 is used as an energy source for **heart pacemakers**. A pacemaker using regular lithium iodide batteries requires changing every 4–5 years. Since plutonium-238 releases energy for many years, its use extends the life of the pacemaker to about 12 years.

4 The age of plant and animal remains is determined by **carbon-14 dating**. Living organisms constantly take in carbon, 0.01% of which is radioactive

carbon-14. When organisms die they stop taking in carbon, and the carbon-14 present begins to undergo radioactive decay – it takes 5600 years to decay to half its original amount. By measuring the percentage of carbon-14 in plant and animal remains, their ages can then be calculated.

5 Radioactive uranium-235 is used to **generate electricity** in nuclear power stations. When an atom of uranium-235 is hit by a neutron, it splits into two smaller atoms, barium and krypton, and releases three neutrons and **heat energy**. These neutrons can then hit three more uranium atoms, causing them to split and release a total of nine neutrons. This sets up a **chain reaction**, which produces an enormous amount of heat energy.

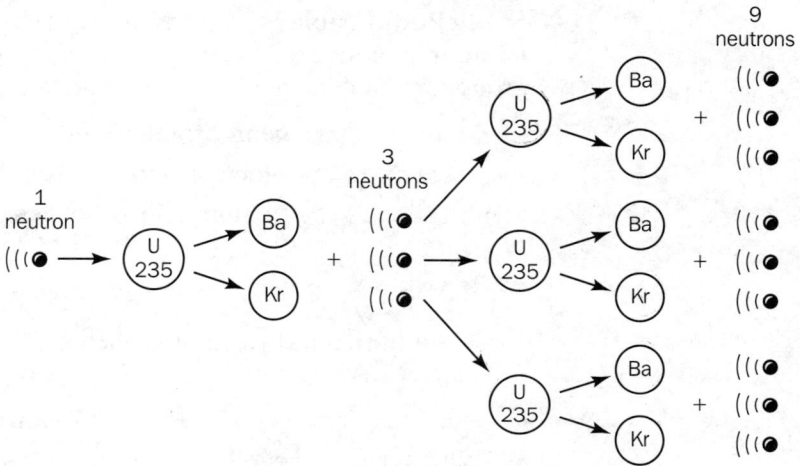

Fig. 3.1 The chain reaction in splitting uranium-235

If this occurs in an uncontrolled way, it results in an **atomic explosion**, e.g. an atom bomb. In a nuclear power station, however, the splitting of uranium-235 is **controlled** so the heat energy can be used to boil water. The steam produced is then used to drive turbines which generate electricity.

The mass of an atom

The **absolute mass (actual mass)** of one atom of an element is difficult to measure owing to size, e.g. the absolute mass of one atom of hydrogen is 1.7×10^{-24} g. The masses of atoms are therefore usually **compared** using their **relative atomic masses**.

Relative atomic mass **(A_r)** *is the average mass of one atom of an element compared with the mass of one atom of* $^{12}_{6}C$ *(carbon-12), the mass of which is taken as exactly 12.00 units.*

Relative atomic masses are not usually whole numbers since they are obtained by taking the **average** relative atomic masses of all the isotopes of the element.

Example

$$\text{Relative atomic mass of chlorine} = \left(\tfrac{75}{100} \times 35\right) + \left(\tfrac{25}{100} \times 37\right) \text{ units}$$
$$= 35.5 \text{ units}$$

This means that the **average mass** of one chlorine atom is $\frac{35.5}{12}$ times the mass of one carbon-12 atom.

4 | *The Periodic Table*

The **Periodic Table** is a classification of all the elements based on **atomic number**. The modern table is composed of horizontal **periods** and vertical **groups** which contain the elements arranged:

- in order of **increasing atomic number**
- in relation to the **electron structure** of their atoms
- in relation to their **chemical properties**.

PERIODS

These are **horizontal rows** of elements. There are **seven** periods numbered **1** to **7**:

- Elements of one period all have the **same number of electron shells**.
- Going **across** a period, each element has **one more proton** and **one more electron** than the previous element.
- Going **across** a period, the chemical properties of elements become **less metallic** and **more non-metallic**.

GROUPS

These are **vertical columns** of elements. There are **eight** groups numbered **I** to **VII** with a final group **0**:

- Each element in a group has the **same number of electrons in its outer shell** – for elements in Groups I to VII, this number is the same as the group number.
- The **common oxidation state (number)** of elements in Groups I to IV, when present in compounds, is the same as the group number. The common oxidation state of elements in Groups V to VII, when present in compounds, is the group number minus 8 (see pp. 62–3). Oxidation state is a measure of the electron control that an atom has when present in a compound, compared to the atom in the pure element.
- Going **down** a group, each element has **one more electron shell** than the previous element.
- Elements in the same group have **similar chemical properties**.
- The **metallic nature** of elements **increases down a group**.
- The **reactivity** of metals **increases down** a group.
- The **reactivity** of non-metals **increases up** a group.

Between Groups II and III are the **transition metals**. All have **two electrons** in their outer shell and most exhibit **variable oxidation states** when in compounds.

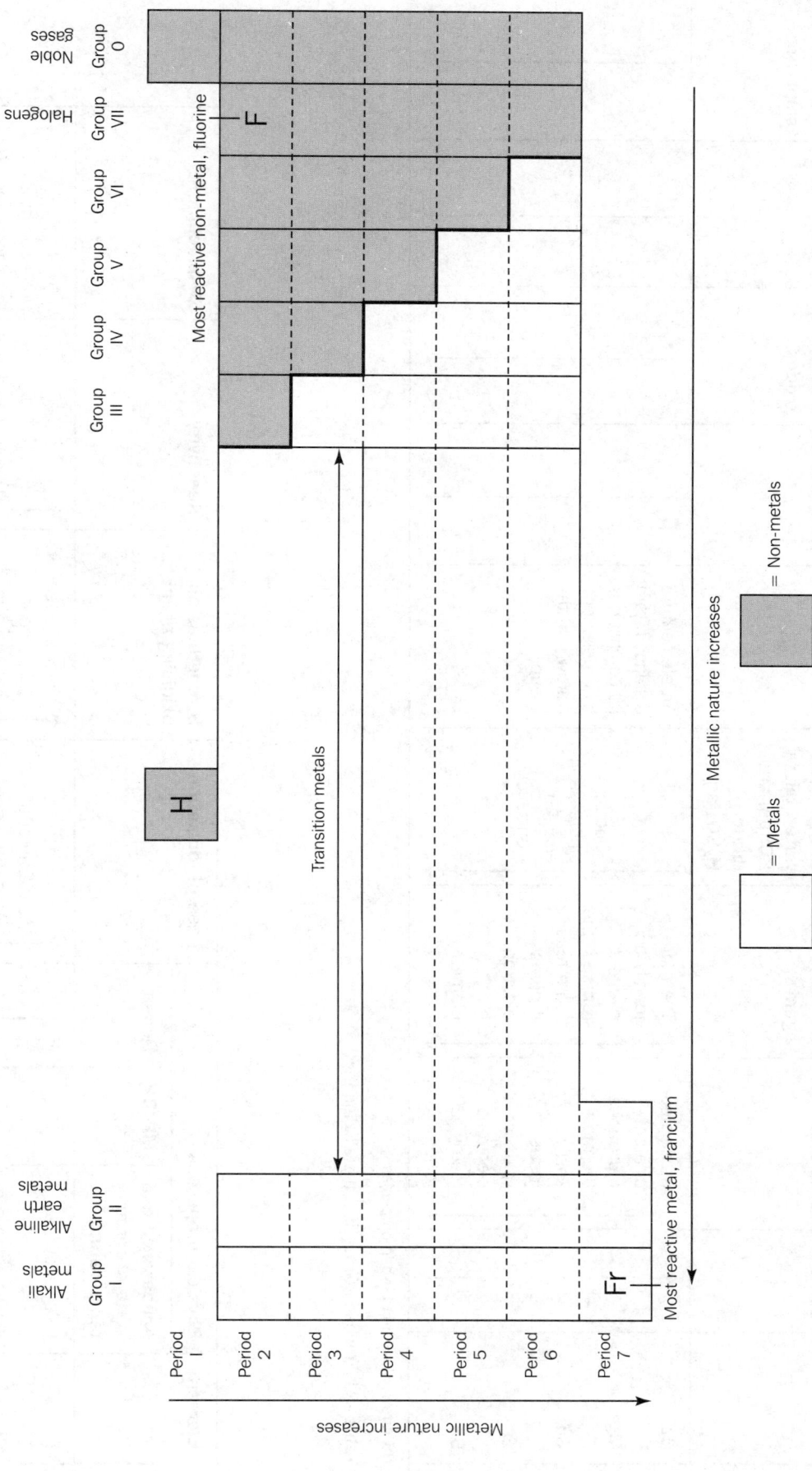

Fig. 4.1 The Periodic Table in outline

Element	Melting point	Ease of ionisation	Strength as a reducing agent	Reactivity with oxygen, dilute water, dilute HCl(aq) and dilute H_2SO_4(aq)	Displacement	Stability of compounds	Basic strength of oxides	Solubility of hydroxides	Solubility of carbonates/sulphates
Be Mg Ca Sr Ba	← (except Mg)	As diameter of atom **increases**, the more easily it **loses** valency electrons (see p. 20) to form ions →	The more readily an element **ionises**, the more readily it **gives** electrons to other elements →	Due to increase in ease of ionisation →	An element is displaced from its compounds by an element **below** it in the group	→	→	→	←

NB Arrows show **increase** in properties.

Table 4.1 Trends in Group II – the alkaline earth metals

Element	Physical properties			Ease of ionisation	Strength as an oxidising agent	Reactivity	Displacement
	Appearance and state at room temperature	MP/BP	Density				
F	Pale yellow gas	→	→	As diameter of atom **decreases**, the more easily it **gains** electrons to form ions ←	The more readily an element **ionises**, the more readily it **removes** electrons from other elements ←	Due to increase in ease of ionisation ←	An element is displaced from its compounds by an element **above** it in the group
Cl	Yellow-green gas						
Br	Red-brown liquid						
I	Grey-black solid						

NB Arrows show **increase** in properties.

Table 4.2 Trends in Group VII – the halogens

Element	Na	Mg	Al	Si	P	S	Cl	Ar
Metallic nature	→ (decreasing across)							
Non-metallic nature	← (increasing across)							
Electrical conductivity	Good conductors			Semi-conductor	Non-conductors			
Ease of ionisation	As number of valency electrons (see p. 20) **decreases**, the fewer electrons the atom has to **lose** to form a cation			As number of valency electrons **increases**, the fewer electrons the atom has to **gain** to form an anion				Does **not** ionise
Reactivity	Due to increased ease of ionisation			Due to increased ease of ionisation				Unreactive
Reducing/oxidising properties	Strength as a reducing agent			Strength as an oxidising agent				Unreactive
Formula of oxide	Na_2O	MgO	Al_2O_3	SiO_2	P_4O_6 P_4O_{10}	SO_2 SO_3	Cl_2O_7	None
Bonding/structure of oxide	Giant ionic lattice		Ionic with some covalent character	Giant atomic lattice	Covalent molecules			—
Nature of oxide	Basic		Amphoteric	Acidic				—
Formula of chloride	NaCl	$MgCl_2$	$AlCl_3$	$SiCl_4$	PCl_3 PCl_5	SCl_2	Cl_2	None
Bonding/structure of chloride	Giant ionic lattice		Ionic with high covalent character	Covalent molecules				—

NB Arrows show **increase** in properties.

Table 4.3 Trends in Period 3

5 Chemical formulae and bonding

Atoms of elements in **Group 0** of the Periodic Table (see p. 17) have **full outer shells** of electrons. They are **stable** and **unreactive** and exist in nature as individual atoms. Atoms of all other elements do not have full outer shells, therefore they are **not stable**. They attempt to gain full outer shells and become stable by:

- **losing electrons** from their outer shell, or
- **gaining electrons** into their outer shell, or
- **sharing electrons** in their outer shell with other atoms.

In doing this, atoms **bond** with each other. There are **three** types of bonding:

- **Ionic or electrovalent bonding** (see p. 21)
- **Covalent bonding** (see p. 24)
- **Metallic bonding** (see p. 27).

Valency of elements

Valency is the number of electrons an atom has to lose, gain or share to attain a stable electronic structure.

Group in Periodic Table	I	II	Transition elements	III	IV	V	VI	VII	0
Valency	1	2	Variable – often 2	3	Usually 4	Usually 3	2	1	0

Table 5.1 The valency of elements

WRITING CHEMICAL FORMULAE OF BINARY COMPOUNDS USING VALENCY

A **chemical formula** represents the **proportions**, by mass, of the different **elements** in a compound.

A **binary compound** is composed of **two** different elements only. Formulae of binary compounds can be written using valencies since both types of atom forming the compound must lose, gain or share the **same number** of electrons.

To write the **chemical formula** of a binary compound:

1 Determine the **valencies** of the two elements present.
2 Write the chemical formula such that the **sums of the valencies** of the two elements are **equal**. Do this by determining the **lowest common multiple** of the two valencies.
3 If a metal is present, always place it **first** in the formula.

Examples

Aluminium oxide

Valencies of elements: Al $= 3$ (Group III)

O $= 2$ (Group VI)

Lowest common multiple of valencies $= 6$

i.e.

Al: $\mathbf{2} \times 3 = 6$

O: $\mathbf{3} \times 2 = 6$

Chemical formula: $\mathbf{Al_2O_3}$

Phosphorus trichloride

Valencies of elements: P $= 3$ (Group V)

Cl $= 1$ (Group VII)

Lowest common multiple of valencies $= 3$

i.e.

P: $\mathbf{1} \times 3 = 3$

Cl: $\mathbf{3} \times 1 = 3$

Chemical formula: $\mathbf{PCl_3}$

A **simpler**, though less scientific method to work out formulae of binary compounds is to 'swap' valencies. Place the valency of the **first** element **after** the symbol of the **second**, and place the valency of the **second** element **after** the symbol of the **first**.

Example

Magnesium nitride

Valencies of elements: Mg $= 2$ (Group II)

N $= 3$ (Group V)

Chemical formula: $\mathbf{Mg_3N_2}$

NB When naming compounds containing an element which can have **more than one valency**, the valency is shown by **Roman numerals** in brackets after the name of the element, e.g.

Copper(I) oxide – valency of Cu $= 1$

valency of O $= 2$

formula: $\mathbf{Cu_2O}$

Copper(II) oxide – valency of Cu $= 2$

valency of O $= 2$

formula: $\mathbf{CuO}$

Ionic or electrovalent bonding

This occurs when a **metal** bonds with a **non-metal**. It involves the **complete transfer** of outer (valency) electrons **from** the metal atom, or atoms, **to** the non-metal atom or atoms. The metal atoms form **positive ions** called **cations** and the non-metal atoms form **negative ions** called **anions**. Ions have full outer shells of electrons.

Examples

Sodium chloride   Formula: NaCl

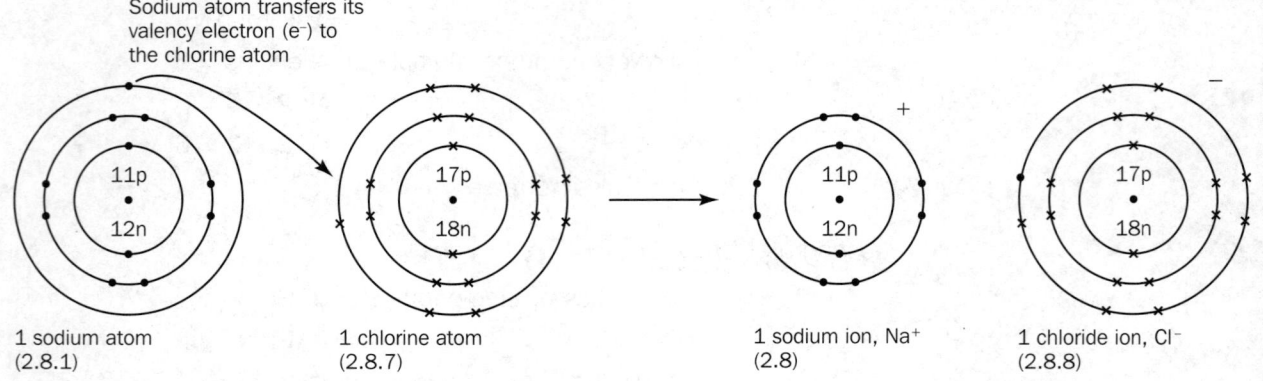

Magnesium fluoride   Formula: MgF_2

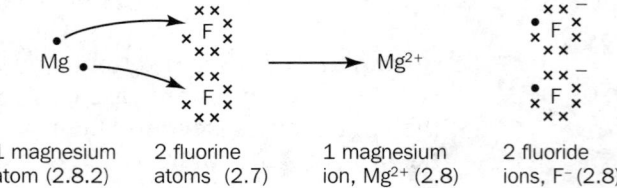

| 1 magnesium | 2 fluorine | 1 magnesium | 2 fluoride |
| atom (2.8.2) | atoms (2.7) | ion, Mg^{2+} (2.8) | ions, F^- (2.8) |

NB   When illustrating the formation of ionic compounds from their elements, it is only necessary to show the **valency electrons**.

THE CRYSTAL STRUCTURE OF IONIC COMPOUNDS

In the **solid** state, ionic compounds exist as **crystals**. Oppositely charged ions are held together in a regular, repeating arrangement throughout each crystal by **strong** electrostatic forces of attraction called **ionic** or **electrovalent bonds**. This forms a three-dimensional structure called a **crystal lattice**.

Example

Sodium chloride

Each Na^+ ion is surrounded by six Cl^- ions, each Cl^- ion is surrounded by six Na^+ ions.

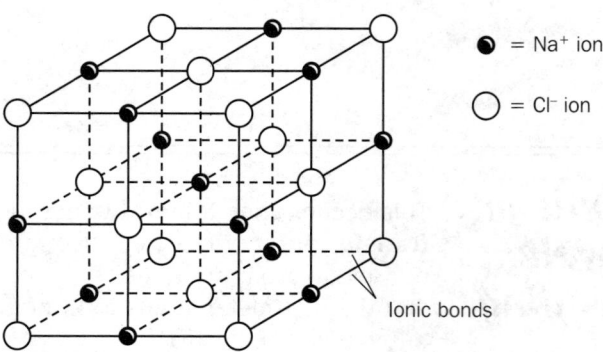

● = Na^+ ion

○ = Cl^- ion

Ionic bonds

Fig. 5.1 The crystal lattice of
             sodium chloride

FORMULAE OF IONIC COMPOUNDS

Ionic compounds may be composed of ions formed from **single atoms** (as shown previously), or ions formed from small **groups of atoms** bonded together called **radicals**, e.g. ammonium, NH_4^+; sulphate, SO_4^{2-}.

Monovalent		Divalent		Trivalent	
Hydrogen	H^+	Magnesium	Mg^{2+}	Iron(III)	Fe^{3+}
Lithium	Li^+	Calcium	Ca^{2+}	Aluminium	Al^{3+}
Sodium	Na^+	Barium	Ba^{2+}		
Potassium	K^+	Iron(II)	Fe^{2+}		
Copper(I)	Cu^+	Copper(II)	Cu^{2+}		
Silver	Ag^+	Zinc	Zn^{2+}		
Ammonium	NH_4^+	Tin(II)	Sn^{2+}		
		Lead(II)	Pb^{2+}		

Table 5.2 Common cations

Monovalent		Divalent		Trivalent	
Fluoride	F^-	Oxide	O^{2-}	Nitride	N^{3-}
Chloride	Cl^-	Sulphide	S^{2-}	Phosphate	PO_4^{3-}
Bromide	Br^-	Carbonate	CO_3^{2-}		
Iodide	I^-	Sulphite	SO_3^{2-}		
Hydride	H^-				
Hydroxide	OH^-	Sulphate	SO_4^{2-}		
Nitrite	NO_2^-	Dichromate(VI)	$Cr_2O_7^{2-}$		
Nitrate	NO_3^-				
Manganate(VII)	MnO_4^-				
Hydrogen sulphate	HSO_4^-				
Hydrogen carbonate	HCO_3^-				
Ethanoate	CH_3COO^-				

Table 5.3 Common anions

NB When **naming anions**, those formed from single atoms are named after the element, with the ending **-ide**, e.g. S^{2-}, sulph**ide**. If **oxygen** is present in a radical, its name is derived from the name of the element combined with oxygen and the ending **-ite** or **ate**, e.g. SO_3^{2-}, sulph**ite**; SO_4^{2-}, sulph**ate**.

When writing formulae of ionic compounds, the **sum** of the **positive charges** must equal the **sum** of the **negative charges**, since both types of atom or radical forming the compound must lose or gain the **same number** of electrons. Formulae of ionic compounds represent the **ratio of ions** present and are called **empirical formulae**.

To write the **empirical formulae** of ionic compounds:

1 Determine the **ions** present.
2 Write the formula such that the **sums** of the positive and negative charges are **equal**. Do this by determining the **lowest common multiple** of the two charges.
3 Place **brackets** around radicals if more than one is present.
4 Always place the metal or ammonium ion **first** in the formula.

Examples

Magnesium nitride

Ions present: Mg^{2+}
N^{3-}

Lowest common multiple of charges = 6

i.e.   Mg: **3** $\times +2 = +6$
N: **2** $\times -3 = -6$

Empirical formula: **Mg_3N_2**

Copper(II) hydroxide

Ions present: Cu^{2+}
OH^-

Lowest common multiple of charges = 2

i.e.   Cu: **1** $\times +2 = +2$
OH: **2** $\times -1 = -2$

Empirical formula: **$Cu(OH)_2$**

Ammonium phosphate

Ions present: NH_4^+
PO_4^{3-}

Lowest common multiple of charges = 3

i.e. NH_4: **3** $\times +1 = +3$
PO_4: **1** $\times -3 = -3$

Empirical formula: **$(NH_4)_3PO_4$**

A **simpler**, though less scientific, method to work out formulae of ionic compounds is to 'swap' charges.

Example

Zinc nitrate

Ions present: Zn^{2+}
NO_3^-

Empirical formula: **$Zn(NO_3)_2$**

Covalent bonding

This occurs when two or more **non-metal** atoms bond. It involves the **sharing** of valency electrons. The shared electrons orbit around the nuclei of both atoms sharing them, forming **strong covalent bonds**. One shared pair forms one covalent bond. Covalent bonding results in the formation of **molecules**.

Formulae of covalent compounds represent **one molecule** of the compound and are called **molecular formulae**.

Examples

Chlorine molecule   Formula: Cl_2

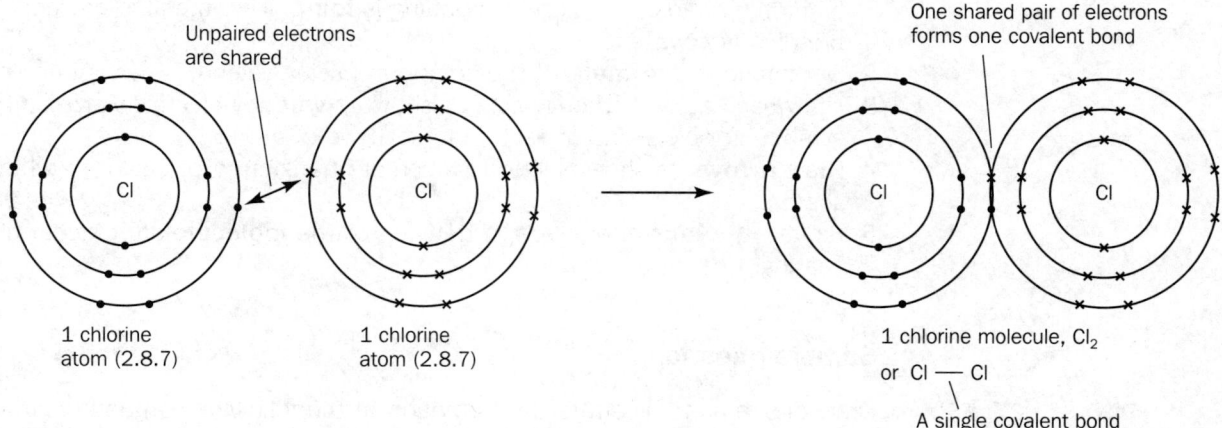

Ammonia molecule   Formula: NH_3

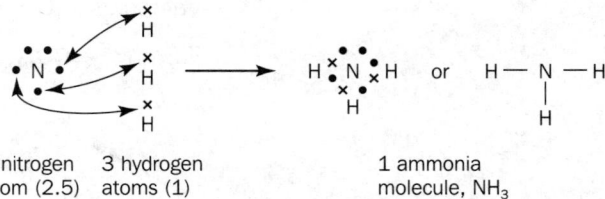

Carbon dioxide molecule   Formula: CO_2

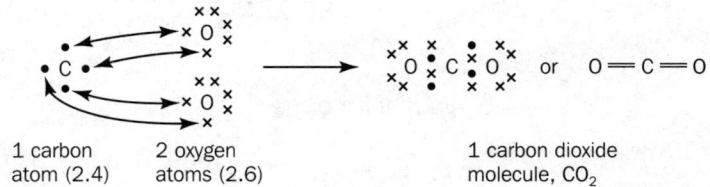

Nitrogen molecule   Formula: N_2

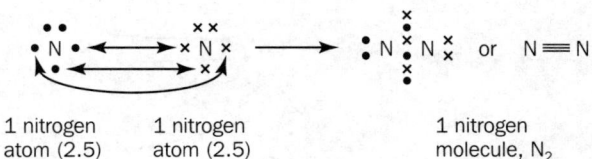

NB   It is only necessary to show **valency electrons**.

Bonding diagrams

DRAWING BONDING DIAGRAMS FOR IONIC AND COVALENT COMPOUNDS

1 Determine if the compound is bonded ionically or covalently – if a metal or ammonium radical is present, bonding is **ionic**; if no metal is present, bonding is **covalent**.
2 Determine the **formula** of the compound using valency, or ions if ionic.
3 Draw each atom in the formula to show its **valency** electrons using different symbols for electrons of each different type of atom.
4 Draw **arrows** to show electrons which are transferred or electrons which are shared.
5 Redraw the **ions** after electron transfer, or the **molecule** after electron sharing.

Sample question

Draw diagrams to illustrate the formation of the following compounds from their elements: **a** tetrachloromethane   **b** aluminium oxide

a Tetrachloromethane

Covalent bonding as **no** metal is present.

Valencies: C = 4; Cl = 1

Formula: **CCl$_4$**

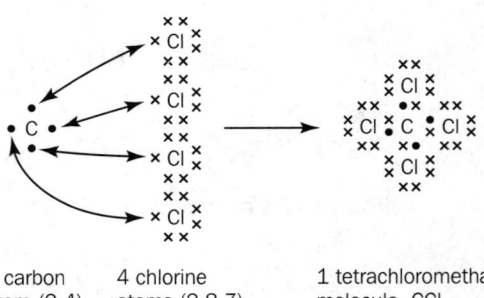

| 1 carbon atom (2.4) | 4 chlorine atoms (2.8.7) | 1 tetrachloromethane molecule, CCl$_4$ |

b Aluminium oxide

Ionic bonding as metal is present.

Ions: Al^{3+}, O^{2-}

Formula: **Al$_2$O$_3$**

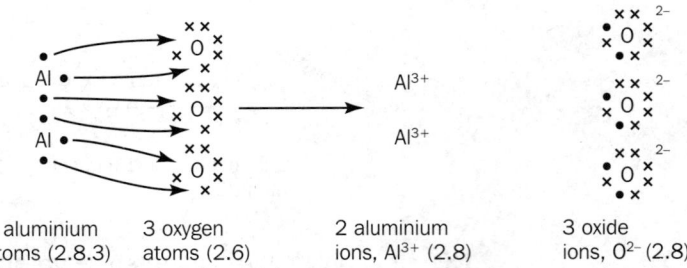

| 2 aluminium atoms (2.8.3) | 3 oxygen atoms (2.6) | 2 aluminium ions, Al^{3+} (2.8) | 3 oxide ions, O^{2-} (2.8) |

Ionic and covalent compounds

Property	Ionic compounds	Simple covalent compounds
Composition	**Ions** held together by strong ionic bonds	**Molecules** with strong covalent bonds between atoms, but weak forces between molecules
State at room temperature	**Crystalline solids** owing to strong ionic bonds holding ions together in a three-dimensional lattice	Most are **liquids** or **gases** owing to weak forces between molecules
Melting and boiling points	**High** – strong ionic bonds require a lot of energy to break	**Low** – weak intermolecular forces require little energy to break
Solubility	Most are soluble in **water** but insoluble in organic (covalent) solvents	Most are soluble in **organic solvents**, e.g. ethanol, but insoluble in water
Electrical conductivity	Conduct electricity when **molten** or **dissolved** in water – ions are free to move	**Do not** conduct electricity in any state – no free ions or electrons are present

Table 5.4 A comparison of the properties of ionic and simple covalent compounds

NB The above properties apply to covalent substances composed of **small molecules**, but not to covalently bonded substances that exist as giant atomic structures, e.g. diamond, graphite (see pp. 28–9).

Metallic bonding

This occurs in **metals**. The metal atoms are packed tightly in **rows** and the valency electrons from each atom are lost to a **'sea' of electrons**. These electrons are **mobile** and bind together the **cations** formed as a result of the atoms losing electrons.

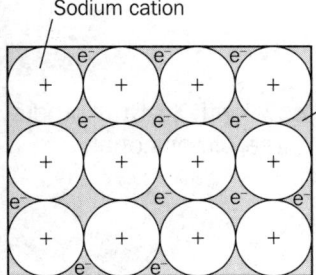

Sodium cation

'Sea' of mobile electrons forms metallic bond

Fig. 5.2 The sodium lattice

PROPERTIES OF METALS

Metals have the following **properties** as a result of their bonding:

1 They are **solid** at room temperature (except mercury): the cations are held together by a strong, cementing 'sea' of electrons.

2 Most have **high** melting and boiling points: the strong metallic bonds require a lot of energy to break.

3 They **conduct electricity**: the mobile electrons can move throughout the metal.

4 They **conduct heat**: heat increases the kinetic energy of mobile electrons – this increase can be passed on through the metal.

5 Most are **malleable** and **ductile**: layers of cations can slip over each other whilst still being held together by the 'sea' of electrons.

6 The structure of solids

Solids can be divided into **four** groups based on structure:

- Ionic crystals
- Atomic (macromolecular or giant molecular) crystals
- Simple molecular crystals
- Metallic crystals.

1 IONIC CRYSTALS

Ionic crystals are composed of an **ionic lattice** in which **cations** and **anions** are held together in a regular, repeating, three-dimensional pattern by strong **ionic bonds**.

Example

All ionic compounds (see p. 22).

2 ATOMIC CRYSTALS

Atomic crystals are composed of an **atomic lattice** in which all the **atoms** are held together in a regular three-dimensional pattern by strong **covalent bonds**. These are also known as **macromolecular** or **giant molecular crystals**.

Examples

Diamond

Diamond is composed of **carbon** atoms each bonded covalently to **four** others arranged in a **tetrahedron** around it. This produces a three-dimensional arrangement of carbon atoms.

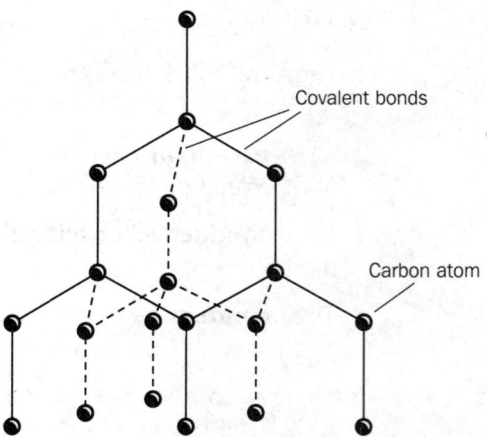

Covalent bonds

Carbon atom

Fig. 6.1 The diamond lattice

Graphite

Graphite is composed of **carbon** atoms each bonded covalently to **three** others to produce **hexagonal** rings of atoms which are arranged in **layers**. The layers are held together by weak forces of attraction. Each carbon atom has a free **mobile** electron.

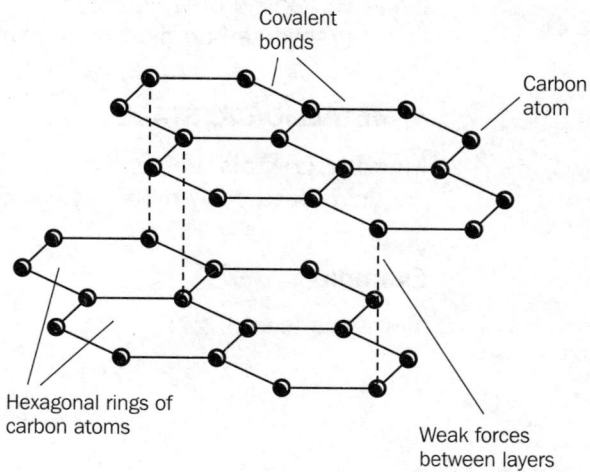

Fig. 6.2 The graphite lattice

Diamond and graphite are **allotropes** of carbon.

Allotropes *are different structural forms of the same element in the same physical state. They have different physical properties but the same chemical properties.*

Physical property	Sodium chloride	Diamond	Graphite
Appearance	**Crystalline solid** – crystals are cubic owing to arrangement of ions (see Fig. 5.1, p. 22)	**Colourless, sparkling solid** – light is reflected around inside the crystal before being reflected out	**Dark grey, opaque solid** – layers are randomly arranged on top of each other so no light is transmitted through
Hardness	**Brittle** – easily split along lines of ions	**Extremely hard** – strong covalent bonds exist throughout structure	**Soft and flaky** – weak bonds exist between layers
Lubricating power	**None** – ions are in layers, but layers are held together by strong ionic bonds	**None** – atoms are bonded covalently throughout crystal	**Extremely good** – layers slip easily over each other owing to weak forces between them
Melting point	**Fairly high** – strong ionic bonds need a lot of energy to break	**Very high** – strong covalent bonds need large amounts of energy to break	**Very high** – strong covalent bonds need large amounts of energy to break
Electrical conductivity	Conducts electricity only when **molten** or **dissolved** in water – ions are then free to move	**Does not** conduct electricity – all electrons are involved in forming covalent bonds	Conducts electricity when **solid** – mobile electrons present, one from each carbon atom

Table 6.1 A comparison of the physical properties of sodium chloride, diamond and graphite

3 SIMPLE MOLECULAR CRYSTALS

Simple molecular crystals are composed of a **molecular lattice** in which **small molecules** are held together by weak **intermolecular forces**.

Examples

Iodine – composed of I_2 molecules.
Sulphur – composed of S_8 molecules.
Ice – composed of H_2O molecules.
Dry ice (solid carbon dioxide) – composed of CO_2 molecules.

4 METALLIC CRYSTALS

Metallic crystals are composed of a **metallic lattice** in which **metal cations** are held together by mobile, **valency electrons**.

Example

All metals (see p. 27).

Chemical equations

A **chemical equation** is a shorthand representation of a chemical reaction. **Reactants** are shown on the **left** and **products** are shown on the **right**, separated by an **arrow**:

$$\text{reactants} \xrightarrow[\text{(e.g. temperature)}]{\text{conditions}} \text{products}$$

Writing equations

1 Write the equation in **words**:

e.g. $\quad \text{zinc hydroxide} + \text{hydrochloric acid} \rightarrow \text{zinc chloride} + \text{water}$

2 Write the correct **formula** of each reactant and product:

$$Zn(OH)_2 + HCl \longrightarrow ZnCl_2 + H_2O$$

3 **Balance** the equation so that there are the **same number** of atoms or ions of each element on each side of the equation. This is done by placing simple whole numbers **in front** of formulae to alter the proportions of reactants and products – **formulae must never be altered**:

$$Zn(OH)_2 + HCl \longrightarrow ZnCl_2 + H_2O$$

Reactants	Products
$Zn = 1$	$Zn = 1$
$O = 2$	$O = 1$
$H = 2 + 1 = 3$	$H = 2$
$Cl = 1$	$Cl = 2$

O, **H** and **Cl** do not balance. **Balance** by placing a **2** in front of **HCl** and a **2** in front of **H$_2$O**:

$$Zn(OH)_2 + \mathbf{2}HCl \longrightarrow ZnCl_2 + \mathbf{2}H_2O$$

i.e.

Reactants	Products
$Zn = 1$	$Zn = 1$
$O = 2$	$O = 2$
$H = 2 + 2 = 4$	$H = 4$
$Cl = 2$	$Cl = 2$

4 Show the **physical state** of each reactant and product by placing a **state symbol** after each formula:

Solid	(s)
Liquid	(l)
Gas	(g)
Aqueous solution	(aq)

$$Zn(OH)_2(\textbf{s}) + 2HCl(\textbf{aq}) \longrightarrow ZnCl_2(\textbf{aq}) + 2H_2O(\textbf{l})$$

Example

Step 1	aluminium	+	hydrochloric acid	$\longrightarrow$	aluminium chloride	+ hydrogen
Step 2	Al	+	HCl	$\longrightarrow$	$AlCl_3$	+ H_2
Step 3	2Al	+	6HCl	$\longrightarrow$	$2AlCl_3$	+ $3H_2$
Step 4	2Al(s)	+	6HCl(aq)	$\longrightarrow$	$2AlCl_3$(aq) +	$3H_2$(g)

USEFUL HINTS FOR BALANCING EQUATIONS

1 Always **begin** with elements which occur in only **one** formula on each side of the equation:

e.g.
$$CuO + NH_4Cl \longrightarrow CuCl_2 + H_2O + NH_3$$

Cl occurs in one formula on each side and does not balance. **Balance** by placing a **2** in front of **NH₄Cl**:

$$CuO + \textbf{2}NH_4Cl \longrightarrow CuCl_2 + H_2O + NH_3$$

Now **N** does not balance. **Balance** by placing a **2** in front of **NH₃**:

$$CuO + 2NH_4Cl \longrightarrow CuCl_2 + H_2O + \textbf{2}NH_3$$

The equation now balances.

2 Treat **radicals** which remain intact during the reaction as **single entities**, never try to balance individual atoms in the radical:

e.g.
$$Pb(NO_3)_2 + NaOH \longrightarrow Pb(OH)_2 + NaNO_3$$

NO₃ does not balance. **Balance** by placing a **2** in front of **NaNO₃**. **OH** does not balance. **Balance** by placing a **2** in front of **NaOH**:

$$Pb(NO_3)_2 + \textbf{2}NaOH \longrightarrow Pb(OH)_2 + 2NaNO_3$$

The equation now balances.

Ionic equations

Ionic equations show only the atoms or ions which actually **take part** in a reaction and, as a result, end up in a **different situation** from the one in which they started, e.g. two ions in solution may join to form a solid precipitate or a covalent compound, or an element in its free state may ionise.

WRITING IONIC EQUATIONS

1 Write the **full**, **balanced** equation:

e.g. $Ba(NO_3)_2(aq) + Na_2SO_4(aq) \longrightarrow BaSO_4(s) + 2NaNO_3(aq)$

2 Rewrite the equation showing the **ions** and their **states**:

$Ba^{2+}(aq)\ 2NO_3^-(aq) + 2Na^+(aq)\ SO_4^{2-}(aq) \longrightarrow Ba^{2+}(s)\ SO_4^{2-}(s) + 2Na^+(aq)\ 2NO_3^-(aq)$

3 Delete any **ions** which remain **unchanged** – NO_3^-**(aq)** and **Na$^+$(aq)** remain free in solution and are therefore unchanged; **Ba^{2+}(aq)** and **SO$_4{}^{2-}$(aq)** change to form a solid precipitate:

$Ba^{2+}(aq)\ \cancel{2NO_3^-(aq)} + \cancel{2Na^+(aq)}\ SO_4^{2-}(aq) \longrightarrow Ba^{2+}(s)\ SO_4^{2-}(s) + \cancel{2Na^+(aq)}\ \cancel{2NO_3^-(aq)}$

4 Rewrite the ionic equation showing only the **ions** which **change**:

$$Ba^{2+}(aq) + SO_4^{2-}(aq) \longrightarrow BaSO_4(s)$$

Example 1
Step 1

$$Na_2CO_3(aq) + 2HCl(aq) \longrightarrow 2NaCl(aq) + H_2O(l) + CO_2(g)$$

Step 2

$2Na^+(aq)\ CO_3^{2-}(aq) + 2H^+(aq)\ 2Cl^-(aq) \longrightarrow 2Na^+(aq)\ 2Cl^-(aq) + H_2O(l) + CO_2(g)$

Step 3 Delete **Na$^+$(aq)** and **Cl$^-$(aq)**:

$\cancel{2Na^+(aq)}\ CO_3^{2-}(aq) + 2H^+(aq)\ \cancel{2Cl^-(aq)} \longrightarrow \cancel{2Na^+(aq)}\ \cancel{2Cl^-(aq)} + H_2O(l) + CO_2(g)$

Step 4 Ionic equation:

$$CO_3^{2-}(aq) + 2H^+(aq) \longrightarrow H_2O(l) + CO_2(g)$$

Example 2
Step 1

$$Zn(s) + 2HNO_3(aq) \longrightarrow Zn(NO_3)_2(aq) + H_2(g)$$

Step 2

$$Zn(s) + 2H^+(aq)\ 2NO_3^-(aq) \longrightarrow Zn^{2+}(aq)\ 2NO_3^-(aq) + H_2(g)$$

Step 3 Delete **NO$_3^-$(aq)**:

$Zn(s) + 2H^+(aq)\ \cancel{2NO_3^-(aq)} \longrightarrow Zn^{2+}(aq)\ \cancel{2NO_3^-(aq)} + H_2(g)$

Step 4 Ionic equation:

$$Zn(s) + 2H^+(aq) \longrightarrow Zn^{2+}(aq) + H_2(g)$$

8 Solutions, solubility, suspensions and colloids

Solutions

A **solution** is a **homogeneous** (same throughout) mixture of two or more substances, one of which is usually a liquid:

- The **solvent** is the substance which does the **dissolving** and is present in the **highest** concentration.
- The **solutes** are the substances which **dissolve** and are present in **lower** concentrations.

See pp. 7–10 for methods used for separating solutions.

Solute	Solvent	Examples
Gas	Gas	Air
Liquid	Liquid	Vinegar
Solid	Solid	Metal alloys, e.g. brass, bronze, steel
Gas	Liquid	Soda water
Solid	Liquid	Sea water

Table 8.1 Types of solution

A **saturated solution** contains as much solute as can be dissolved at a given temperature, in the presence of undissolved solute.

Solubility

Solubility *is the mass of solute which will saturate 100 g of solvent at a given temperature.*

In general:

- The solubility of a **solid** in a liquid **increases** as temperature increases.
- The solubility of a **gas** in a liquid **decreases** as temperature increases.

SOLUBILITY CURVES

When solubility is plotted against temperature, a **solubility curve** is obtained. For example, see Fig. 8.1.

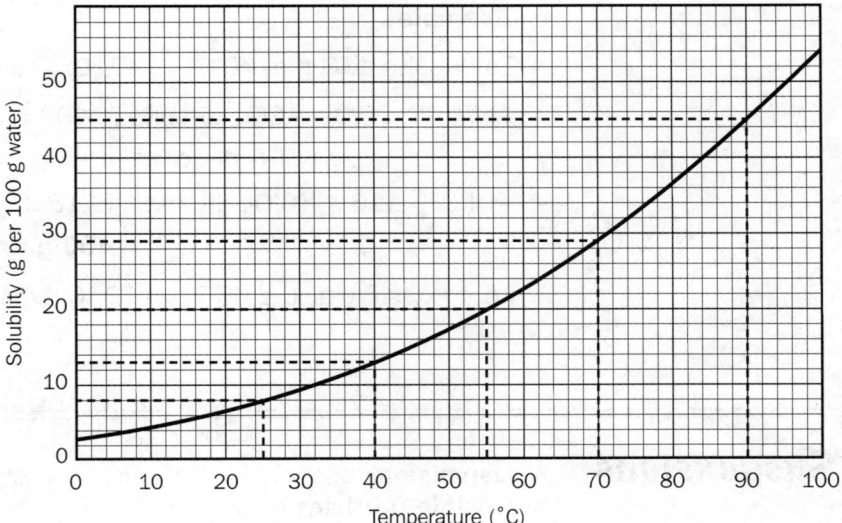

Fig. 8.1 Solubility curve for potassium chlorate(v), KClO₃, in water

Solubility curves can be used to obtain the following information:

1 The **solubility** of a solid at any **given temperature**.

Example

At **25** °C, solubility of KClO₃ = **8.0 g per 100 g water**

2 The **mass of solute** which must be **added** to resaturate a solution if its temperature is **increased**.

Example

If the temperature is increased from **40 °C** to **55 °C**:

Solubility of KClO₃ at 40 °C = 13.0 g per 100 g water
Solubility of KClO₃ at 55 °C = 20.0 g per 100 g water
and   20.0 − 13.0 g = **7.0 g**

i.e. **7.0 g** of KClO₃ must be added to a solution containing 100 g of water at 40 °C to form a saturated solution at 55 °C.

3 The **mass of solute** which would **crystallise out** of a saturated solution if its temperature is **decreased**.

Example

If the temperature is decreased from **90 °C** to **70 °C**:

Solubility of KClO₃ at 90 °C = 45.0 g per 100 g water
Solubility of KClO₃ at 70 °C = 29.0 g per 100 g water
and   45.0 − 29.0 g = **16.0 g**

i.e. **16.0 g** of KClO₃ would crystallise out if a solution containing 100 g of water that is saturated at 90 °C is cooled to 70 °C.

4 The **minimum mass of solvent** required to dissolve a **fixed mass of solute** at a given temperature.

Example

To dissolve **116 g** of $KClO_3$ at **70 °C**:

At 70 °C, 29.0 g $KClO_3$ dissolve in 100 g water.

$\therefore$ 1 g $KClO_3$ dissolves in $\frac{100}{29}$ g water

and   **116 g $KClO_3$** dissolve in $116 \times \frac{100}{29}$ g water

$= $ **400 g water**

i.e. to dissolve 116 g $KClO_3$ at 70 °C requires **400 g** of water.

Suspensions

A **suspension** consists of a substance, usually a liquid, which contains **minute**, but **visible particles** held floating in the substance. If left undisturbed, the particles eventually **settle**.

A suspension appears **cloudy**. The particles in a suspension can be removed using **filter paper**.

Examples

- **Dust** in **air**: a suspension of a solid in a gas.
- **Mist** in **air**: a suspension of a liquid in a gas.
- **Powdered chalk** in **water**: a suspension of a solid in a liquid.

Colloids

A **colloid** consists of a substance, usually a liquid, which contains **macromolecules** or **aggregates of macromolecules** held floating in the substance. The particles cannot be seen even with a microscope. If left undisturbed, the particles **do not settle**.

A colloid **scatters light**. The particles in a colloid can pass through filter paper, but can be removed by a **parchment membrane (permeable membrane)**.

Examples

- **Smoke** in **air**: a colloid of a solid in a gas.
- **Fog** or **aerosol sprays** in **air**: colloids of a liquid in a gas.
- **Starch** dissolved in **water**: a colloid of a solid in a liquid.
- **Emulsions**, e.g. milk, mayonnaise: colloids of a liquid (oil) in a liquid.

Fig. 8.2 A comparison of the size of particles in solutions, colloids and suspensions

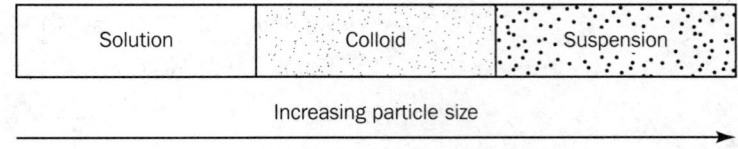

9 Acids, bases and salts

Acids

Acids in their **pure anhydrous** form are composed of **covalent molecules**. However, acids dissolve in water to form solutions which contain **ions**. These solutions are described as **acidic**.

Acids may be defined in **three** ways:

1 An *acid* is a substance which contains *hydrogen* which can be replaced directly or indirectly by a metal to form a salt,

e.g. $$Zn(s) + 2HCl(aq) \longrightarrow ZnCl_2(aq) + H_2(g)$$

The zinc replaces the hydrogen to form the salt, zinc chloride.

2 An *acid* is a substance which forms *hydroxonium ions* when dissolved in water.

When an acid dissolves in water it initially forms **hydrogen ions**. Each hydrogen ion then becomes associated with a water molecule to form a hydroxonium ion,

e.g. $$HCl(g) + water \longrightarrow H^+(aq) + Cl^-(aq)$$
$$H^+(aq) + H_2O(l) \longrightarrow \underset{\text{hydroxonium ion}}{H_3O^+(aq)}$$

Overall reaction: $$HCl(g) + H_2O(l) \longrightarrow H_3O^+(aq) + Cl^-(aq)$$

For simplicity, it is usual to represent $H_3O^+(aq)$ as $H^+(aq)$.

3 An *acid* is a *proton donor*.

When a hydrogen atom loses an electron to form an H^+ ion, it becomes a hydrogen nucleus which contains a single proton. Hydrogen ions are, therefore, **protons** and acids can donate these to other substances. For example, when an acid dissolves in water it **donates** its hydrogen ions (protons) to the water molecules,

e.g. $$HCl(g) + H_2O(l) \longrightarrow H_3O^+(aq) + Cl^-(aq)$$

GENERAL PROPERTIES OF AQUEOUS ACIDS

The properties of acids in their pure form are quite different from those of their aqueous solutions. It is the **H^+(aq) ions** that cause the **acidic properties** listed below and these are only formed in the presence of **water**.

1 Acids have a **sour** taste.
2 Acids are **corrosive**.
3 Acids turn **litmus** from blue to **red** (see Table 9.1, p. 40).
4 Acids conduct electricity, i.e. they are **electrolytes**.

5 Acids react with **bases** to form a **salt** and **water** only:

$$\text{base} + \text{acid} \longrightarrow \text{salt} + \text{water}$$

e.g.

$$Mg(OH)_2(s) + 2HCl(aq) \longrightarrow MgCl_2(aq) + 2H_2O(l)$$

$$CuO(s) + H_2SO_4(aq) \longrightarrow CuSO_4(aq) + H_2O(l)$$

Ionically: $OH^-(s \text{ or } aq) + H^+(aq) \longrightarrow H_2O(l)$

or $O^{2-}(s) + 2H^+(aq) \longrightarrow H_2O(l)$

This type of reaction is called a **neutralisation** reaction.

6 Acids react with **carbonates** or **hydrogencarbonates** to form a **salt**, **water** and **carbon dioxide**:

$$\begin{array}{c}\text{carbonate or}\\ \text{hydrogencarbonate}\end{array} + \text{acid} \longrightarrow \text{salt} + \text{water} + \text{carbon dioxide}$$

e.g. $CaCO_3(s) + 2HCl(aq) \longrightarrow CaCl_2(aq) + H_2O(l) + CO_2(g)$

$NaHCO_3(aq) + HNO_3(aq) \longrightarrow NaNO_3(aq) + H_2O(l) + CO_2(g)$

Ionically:

$CO_3{}^{2-}(s \text{ or } aq) + 2H^+(aq) \longrightarrow H_2O(l) + CO_2(g)$

or $HCO_3{}^-(aq) + H^+(aq) \longrightarrow H_2O(l) + CO_2(g)$

7 Acids, except nitric acid, react with the **reactive metals** (iron and above in the reactivity series – see Table 17.1, p. 90) to form a **salt** and **hydrogen**:

$$\text{reactive metal} + \text{acid} \longrightarrow \text{salt} + \text{hydrogen}$$

e.g. $Mg(s) + 2HCl(aq) \longrightarrow MgCl_2(aq) + H_2(g)$

Ionically $Mg(s) + 2H^+(aq) \longrightarrow Mg^{2+}(aq) + H_2(g)$

Since nitric acid is an oxidising agent it releases **oxides of nitrogen**, e.g. nitrogen dioxide, NO_2, and not hydrogen when reacted with metals.

THE BASICITY (PROTICITY) OF ACIDS

The **basicity of an acid** is the number of **moles of H^+ ions** produced per mole of acid.

1 **Monobasic acids** produce **one** mole of H^+ ions per mole of acid,

e.g.

$HCl(aq) \longrightarrow H^+(aq) + Cl^-(aq)$
1 mole   1 mole

$HNO_3(aq) \longrightarrow H^+(aq) + NO_3{}^-(aq)$
1 mole   1 mole

Monobasic acids form **normal salts** only (see p. 42).

2 **Dibasic acids** produce **two** moles of H^+ ions per mole of acid,

e.g.

$H_2SO_4(aq) \longrightarrow 2H^+(aq) + SO_4{}^{2-}(aq)$
1 mole   2 moles

Dibasic acids can form **normal salts** and **acid salts** (see p. 42).

3 **Tribasic acids** produce **three** moles of H^+ ions per mole of acid,

$H_3PO_4(aq) \longrightarrow 3H^+(aq) + PO_4{}^{3-}(aq)$
1 mole   3 moles

Tribasic acids can form **normal salts** and **acid salts**.

ACID ANHYDRIDES

An **acid anhydride** is an **acidic oxide** of a non-metal which reacts with water to form an **acid**,

e.g.
$$CO_2(g) + H_2O(l) \rightleftharpoons H_2CO_3(aq)$$
$$SO_2(g) + H_2O(l) \rightleftharpoons H_2SO_3(aq)$$
$$SO_3(g) + H_2O(l) \longrightarrow H_2SO_4(aq)$$

ACIDS IN DAILY LIFE

1 **Antacids** contain sodium hydrogencarbonate and usually **citric** or **tartaric acid**. The sodium hydrogencarbonate **neutralises** stomach acid, but has an unpleasant taste. When water is added, some of the sodium hydrogencarbonate reacts with the acid producing **carbon dioxide**. The carbon dioxide causes the antacid to **fizz** and be **less unpleasant tasting**. When taken as a drink, the remaining sodium hydrogencarbonate neutralises the stomach acid.

2 **Baking powder** is **tartaric acid** combined with sodium hydrogencarbonate. Heat causes the sodium hydrogencarbonate to decompose releasing **carbon dioxide**:

$$2NaHCO_3(s) \longrightarrow Na_2CO_3(s) + H_2O(l) + CO_2(g)$$

The tartaric acid then reacts with the sodium carbonate in the presence of moisture forming more **carbon dioxide**. The carbon dioxide forms bubbles in the cake causing it to **rise**.

3 **Fire extinguishers** contain **carbon dioxide** under pressure. When pressure is reduced by opening the valve, the carbon dioxide is released. Carbon dioxide is **non-flammable** and, due to its **high density**, it smothers the fire.

Bases

A **base** is chemically opposite to an acid and may be defined in **two** ways:

1 A *base is a substance which will react with an* **acid** *to form a salt and water only.*
2 A *base is a* **proton** (H$^+$ ion) **acceptor**.

For example, when a metal hydroxide reacts with an acid, the hydroxide ions **accept** hydrogen ions (protons) from the acid forming water:

$$\underset{\text{(from base)}}{OH^-(s \text{ or } aq)} + \underset{\text{(from acid)}}{H^+(aq)} \longrightarrow H_2O(l)$$

Bases include **ammonia** and most **metal oxides** and **hydroxides**, e.g. magnesium oxide, MgO; copper(II) hydroxide, Cu(OH)$_2$.

Alkalis

An **alkali** is a base which dissolves in, or reacts with, water to form a solution which contains **OH$^-$ ions**. The solution is described as **alkaline**,

e.g.
$$NaOH(s) + water \longrightarrow Na^+(aq) + OH^-(aq)$$
$$Na_2O(s) + H_2O(l) \longrightarrow 2Na^+(aq) + 2OH^-(aq)$$
$$NH_3(g) + H_2O(l) \rightleftharpoons NH_4^+(aq) + OH^-(aq)$$

Most bases are **insoluble**, and are therefore **not** alkalis.

GENERAL PROPERTIES OF AQUEOUS ALKALIS

1 Alkalis have a **bitter** taste.
2 Alkalis are **soapy** to touch.
3 Alkalis are **corrosive**.
4 Alkalis turn **litmus** from red to **blue** (see Table 9.1, below).
5 Alkalis conduct electricity, i.e. they are **electrolytes**.
6 Alkalis react with solutions containing **metal ions** (except potassium and sodium) to form insoluble **precipitates**,

e.g. $CuSO_4(aq) + 2NaOH(aq) \longrightarrow Cu(OH)_2(s) + Na_2SO_4(aq)$

Ionically: $M^{n+}(aq) + nOH^-(aq) \longrightarrow M(OH)_n(s)$

7 Alkalis react with **zinc** and **aluminium** forming a **salt** and **hydrogen**:

$$\begin{matrix} \textbf{zinc or} \\ \textbf{aluminium} \end{matrix} + \textbf{alkali} \longrightarrow \textbf{salt} + \textbf{hydrogen}$$

e.g. $Zn(s) + 2NaOH(aq) \longrightarrow \underset{\text{sodium zincate}}{Na_2ZnO_2(aq)} + H_2(g)$

8 Alkalis react with **ammonium salts** forming a **salt**, **water** and **ammonia**:

$$\textbf{alkali} + \begin{matrix} \textbf{ammonium} \\ \textbf{salt} \end{matrix} \longrightarrow \textbf{salt} + \textbf{water} + \textbf{ammonia}$$

e.g. $NaOH(aq) + NH_4Cl(aq) \longrightarrow NaCl(aq) + H_2O(l) + NH_3(g)$

Ionically: $OH^-(aq) + NH_4^+(aq) \longrightarrow H_2O(l) + NH_3(g)$

Insoluble bases react slightly in this way.

9 **All bases** react with **acids** forming a **salt** and **water** only (see p. 38).

Recognising acids and alkalis

Acids and **alkalis** are recognised using **indicators**. An indicator is a substance which has one colour when mixed with an acidic solution and a different colour when mixed with an alkaline solution.

Indicator	Colour in acidic solution	Colour in alkaline solution
Litmus	Red	Blue
Phenolphthalein	Colourless	Pink
Methyl orange	Pink/red	Yellow
Screened methyl orange	Red	Green
Bromothymol blue	Yellow	Blue

Table 9.1 Some common indicators

The strength of acids and alkalis

1 **Strong acids** and **strong alkalis** are **fully ionised** when they dissolve in water, i.e. their solutions contain a high concentration of H^+ ions or OH^- ions, respectively. They are **strong electrolytes** (see p. 68),

e.g. $H_2SO_4(aq) \longrightarrow 2H^+(aq) + SO_4^{2-}(aq)$

$NaOH(aq) \longrightarrow Na^+(aq) + OH^-(aq)$

2 **Weak acids** and **weak alkalis** are only **partially ionised** when they dissolve in water, i.e. their solutions contain a low concentration of H^+ ions or OH^- ions, respectively. They are **weak electrolytes**,

e.g.

$$CH_3COOH(aq) \rightleftharpoons CH_3COO^-(aq) + H^+(aq)$$

$$NH_3(g) + H_2O(l) \rightleftharpoons NH_4^+(aq) + OH^-(aq)$$

	Strong		**Weak**	
Acids	Hydrochloric acid	HCl	Nitrous acid	HNO_2
	Nitric acid	HNO_3	Sulphurous acid	H_2SO_3
	Sulphuric acid	H_2SO_4	Carbonic acid	H_2CO_3
	Phosphoric acid	H_3PO_4	Ethanoic acid	CH_3COOH
Alkalis	Potassium hydroxide	KOH	Ammonia solution	$NH_3(aq)$
	Sodium hydroxide	NaOH	Calcium hydroxide	$Ca(OH)_2$

Table 9.2 The common acids and alkalis and their strengths

THE pH SCALE

The **strength** of an acid or alkali is measured on the **pH scale** using **universal indicator**.

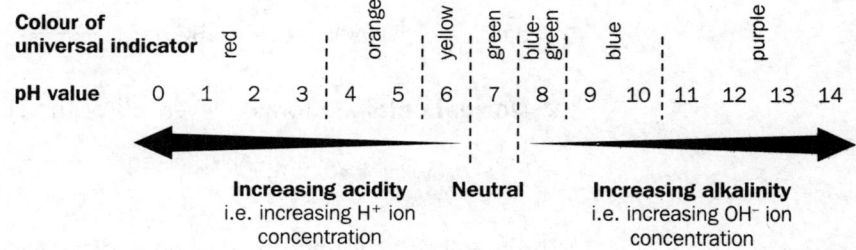

Fig. 9.1 The pH scale

Amphoteric oxides and hydroxides

An **amphoteric substance** is a substance which can react with both **acids** and **strong alkalis**.

$$\begin{matrix} \text{strong} \\ \text{alkali} \end{matrix} + \begin{matrix} \text{amphoteric oxide} \\ \text{or hydroxide} \end{matrix} \longrightarrow \text{salt} + \text{water}$$

$$\begin{matrix} \text{amphoteric oxide} \\ \text{or hydroxide} \end{matrix} + \text{acid} \longrightarrow \text{salt} + \text{water}$$

Examples

- Aluminium hydroxide reacts with the strong alkali, sodium hydroxide:

$$NaOH(aq) + Al(OH)_3(s) \longrightarrow NaAlO_2(aq) \text{sodium aluminate} + 2H_2O(l)$$

- Aluminium hydroxide reacts with hydrochloric acid:

$$Al(OH)_3(s) + 3HCl(aq) \longrightarrow AlCl_3(aq) + 3H_2O(l)$$

The following substances are amphoteric:

Aluminium hydroxide, **Al(OH)$_3$**   Aluminium oxide, **Al$_2$O$_3$**

Zinc hydroxide, **Zn(OH)$_2$**   Zinc oxide, **ZnO**

Lead(II) hydroxide, **Pb(OH)$_2$**   Lead(II) oxide, **PbO**

Salts

A **salt** is a compound formed when some or all of the hydrogen ions in an acid are replaced by **metal** or **ammonium ions**; these ions may be supplied by the metal itself, a carbonate, a hydrogen carbonate or a base (see p. 38).

Salts are **ionic** and contain at least one **metallic** or **ammonium cation**, and at least one **anion (acid radical)** from the **acid**.

TYPES OF SALTS

There are **two** types of salts:

1 **Acid salts** are formed when only **some** of the **H$^+$ ions** are replaced. Acid salts, therefore, contain some H$^+$ ions from the original acid. Only **dibasic** and **tribasic** acids can form acid salts,

e.g. $NaOH(aq) + H_2SO_4(aq) \longrightarrow NaHSO_4(aq) + H_2O(l)$
          (acid salt)

2 **Normal salts** are formed when **all** of the **H$^+$ ions** are replaced,

e.g. $2NaOH(aq) + H_2SO_4(aq) \longrightarrow Na_2SO_4(aq) + 2H_2O(l)$
          (normal salt)

The type of salt formed by dibasic and tribasic acids depends on the **quantity** of acid used.

Acid	Salts formed	Anion present	Example
Hydrochloric acid	Chlorides	Cl^-	$NaCl$
Nitric acid	Nitrates	NO_3^-	$NaNO_3$
Ethanoic acid	Ethanoates	CH_3COO^-	CH_3COONa
Sulphuric acid	Hydrogen sulphates (acid salts)	HSO_4^-	$NaHSO_4$
	Sulphates	SO_4^{2-}	Na_2SO_4
Carbonic acid	Hydrogen carbonates (acid salts)	HCO_3^-	$NaHCO_3$
	Carbonates	CO_3^{2-}	Na_2CO_3
Phosphoric acid	Dihydrogen phosphates (acid salts)	$H_2PO_4^-$	NaH_2PO_4
	Hydrogen phosphates (acid salts)	HPO_4^{2-}	Na_2HPO_4
	Phosphates	PO_4^{3-}	Na_3PO_4

Table 9.3 Salts formed by common acids

WATER OF CRYSTALLISATION

Most salts contain a fixed amount of water in their crystal lattice called **water of crystallisation**. This water is essential to the **shape** and sometimes the **colour** of the crystals. If removed, the salt becomes **anhydrous** and the colour and shape of the crystals may change,

e.g. $CuSO_4 . 5H_2O(s) \xrightarrow{heat} CuSO_4(s) + 5H_2O(g)$
blue − hydrated   **white** − anhydrous

Methods of preparing salts

When preparing salts, the following must be considered:

1 The **solubility** of the **salt**.
2 The **solubility** of the **starting materials**.
3 The **hydration** of the **salt**.

	Salt or base	Solubility in water	Exceptions
Salts	Chlorides	Soluble	**AgCl** is insoluble; **PbCl$_2$** is soluble in hot water, insoluble in cold water
	Sulphates	Soluble	**PbSO$_4$** and **BaSO$_4$** are insoluble; **CaSO$_4$** is slightly soluble
	Nitrates	Soluble	None
	Carbonates	Insoluble	**K$_2$CO$_3$**, **Na$_2$CO$_3$** and **(NH$_4$)$_2$CO$_3$** are soluble
	Potassium, sodium and ammonium salts	Soluble	None
Bases	Oxides	Insoluble	**K$_2$O**, **Na$_2$O** and **CaO** react with water forming the corresponding hydroxide
	Hydroxides	Insoluble	**KOH**, **NaOH** and **NH$_4$OH** are soluble; **Ca(OH)$_2$** is slightly soluble

Table 9.4 The solubility of salts and bases

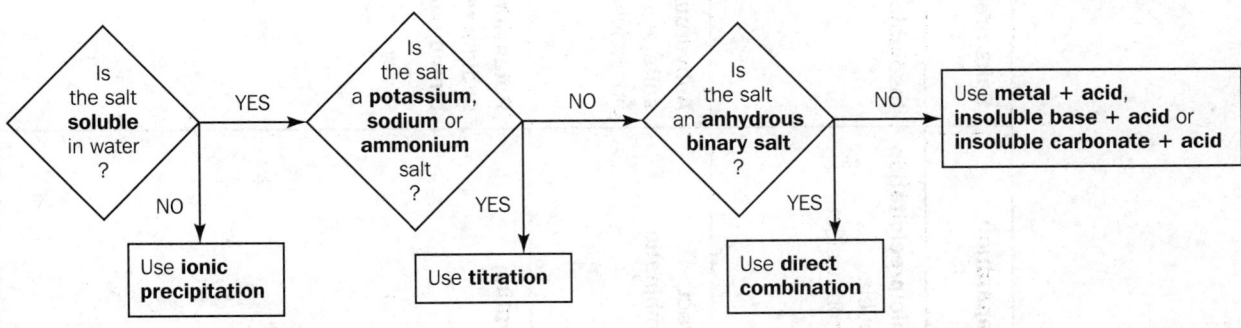

Fig. 9.2 A flow diagram for the preparation of salts (see Table 9.5 for the detailed methods).

Preparation	Salts prepared	Starting materials	Method	Examples	Starting materials and equation for examples
Ionic precipitation (double decomposition)	**Insoluble salts**	**Two solutions** – one provides cations, one provides anions	• **Mix** solutions, precipitate forms • **Filter** to separate precipitate • **Wash** precipitate with distilled water • **Dry** precipitate	$PbCl_2$	**$Pb(NO_3)_2(aq)$** to provide **Pb^{2+} ions**; **NaCl(aq)** to provide **Cl^- ions:** $Pb^{2+}(aq) + 2Cl^-(aq) \longrightarrow PbCl_2(s)$
Direct combination	**Anhydrous binary salts**, e.g. chlorides and sulphides	Appropriate **metal** to provide cations. Appropriate **non-metal** to provide anions	• **Heat** the two elements together – in case of chlorides, pass chlorine gas over heated metal	$FeCl_3$	**Fe(s)** to provide **Fe^{3+} ions**; **$Cl_2(g)$** to provide **Cl^- ions:** $2Fe(s) + 3Cl_2(g) \xrightarrow{heat} 2FeCl_3(s)$
Titration	**Potassium, sodium** or **ammonium salts**	**Two solutions** – one an **alkali** or **carbonate** to provide K^+, Na^+ or NH_4^+ ions; one an **acid** to provide anions	• Determine volume of acid, from a **burette**, needed to neutralise a **fixed volume** of alkali or carbonate solution, measured in a **pipette**, by repeated **titrations** using an **indicator** until two volumes of acid are the same • **Add** volume of acid determined above to fixed volume of alkali or carbonate solution without indicator • **Evaporate** some of the water • Leave to **crystallise** • **Rinse** and **dry** crystals	Na_2SO_4 $NaHSO_4$	**NaOH(aq)** to provide **Na^+ ions**; **$H_2SO_4(aq)$** to provide **SO_4^{2-} ions:** $2NaOH(aq) + H_2SO_4(aq)$ $\longrightarrow Na_2SO_4(aq) + 2H_2O(l)$ [1] **NaOH(aq)** to provide **Na^+ ions**; **$H_2SO_4(aq)$** to provide **HSO_4^- ions:** $NaOH(aq) + H_2SO_4(aq)$ $\longrightarrow NaHSO_4(aq) + H_2O(l)$ [2] Comparing equations [1] and [2], **half** the volume of **alkali** or **twice** the volume of **acid** is needed to make an **acid salt**

(continued)

Table 9.5 Methods of preparing salts

Preparation	Salts prepared	Starting materials	Method	Examples	Starting materials and equation for examples
Metal + acid	**Soluble salts** of the reactive metals, magnesium, aluminium, zinc and iron	Appropriate **metal** to provide cations. Appropriate **acid** to provide anions	• Add **metal** to **acid**, warming if necessary, until effervescence stops and excess metal is present • **Filter** to remove excess metal, collect filtrate • **Evaporate** some water • Leave to **crystallise** • **Rinse** and **dry** crystals	$MgCl_2$	**Mg(s)** to provide **Mg^{2+} ions;** **HCl(aq)** to provide **Cl^- ions:** $Mg(s) + 2HCl(aq) \longrightarrow MgCl_2(aq) + H_2(g)$
Insoluble base + acid	**Soluble salts** except sodium, potassium and ammonium salts	Appropriate **insoluble base** to provide cations. Appropriate **acid** to provide anions	• Add **base** to **acid**, stirring and warming, until excess base is present • **Filter** to remove excess base, collect filtrate • **Evaporate** some water • Leave to **crystallise** • **Rinse** and **dry** crystals	$CuSO_4$	**CuO(s)** to provide **Cu^{2+} ions;** **H_2SO_4(aq)** to provide **SO_4^{2-} ions:** $CuO(s) + H_2SO_4(aq) \longrightarrow CuSO_4(aq) + H_2O(l)$
Insoluble carbonate + acid	**Soluble salts** except sodium, potassium and ammonium salts	Appropriate **insoluble carbonate** to provide cations. Appropriate **acid** to provide anions	• Add **carbonate** to **acid** until effervescence stops and excess carbonate is present • **Filter** to remove excess carbonate, collect filtrate • **Evaporate** some water • Leave to **crystallise** • **Rinse** and **dry** crystals	$Ca(NO_3)_2$	**$CaCO_3$(s)** to provide **Ca^{2+} ions;** **HNO_3(aq)** to provide **NO_3^- ions:** $CaCO_3(s) + 2HNO_3(aq) \longrightarrow Ca(NO_3)_2(aq) + H_2O(l) + CO_2(g)$

Table 9.5 *Continued*

45

10 The mole and chemical calculations

Relative atomic, molecular and formula masses

- **Relative atomic mass (A_r)** *is the average mass of one atom of an element compared with the mass of one atom of carbon-12, the mass of which is taken as exactly 12.00 units.*

- **Relative molecular mass (M_r)** *is the average mass of one molecule of an element or compound compared with the mass of one atom of carbon-12, the mass of which is taken as exactly 12.00 units.*

  This applies to **molecular** substances.

- **Relative formula mass** *is the average mass of one formula unit of a compound compared with the mass of one atom of carbon-12, the mass of which is taken as exactly 12.00 units.*

  This applies to **ionic** compounds.

CALCULATING RELATIVE ATOMIC, MOLECULAR AND FORMULA MASSES

Relative atomic mass, taken to the nearest whole number, is usually the same as the **mass number** of the isotope which occurs in the highest proportion, e.g.

$$\text{Magnesium:} \quad \mathbf{Mg = 24}$$
$$\text{Oxygen:} \quad \mathbf{O = 16}$$
$$\text{Copper:} \quad \mathbf{Cu = 64}$$

Relative molecular mass and **relative formula mass** are calculated by **adding** together the relative atomic masses of all **atoms** in the molecule or all **ions** in the formula unit.

Examples

Relative molecular mass

Chlorine: $\mathbf{Cl_2} = 2 \times 35.5$
$\qquad = \mathbf{71}$

Water: $\mathbf{H_2O} = (2 \times 1) + 16$
$\qquad = 2 + 16$
$\qquad = \mathbf{18}$

Glucose: $C_6H_{12}O_6 = (6 \times 12) + (12 \times 1) + (6 \times 16)$
$$= 72 + 12 + 96$$
$$= \textbf{180}$$

Relative formula mass

Aluminium oxide: $Al_2O_3 = (2 \times 27) + (3 \times 16)$
$$= 54 + 48$$
$$= \textbf{102}$$

Ammonium sulphate: $(NH_4)_2SO_4 = (2 \times 14) + (8 \times 1) + 32 + (4 \times 16)$
$$= 28 + 8 + 32 + 64$$
$$= \textbf{132}$$

The mole

It has been found that **12.00 g** of carbon-12 contain $\mathbf{6.02 \times 10^{23}}$ **carbon atoms**. This number is called the **Avogadro number** or **Avogadro constant**. Similarly:

24 g of magnesium (Mg) contain $\mathbf{6.02 \times 10^{23}}$ **Mg atoms**.

18 g of water (H_2O) contain $\mathbf{6.02 \times 10^{23}}$ $\mathbf{H_2O}$ **molecules**

102 g of aluminium oxide (Al_2O_3) contain $\mathbf{6.02 \times 10^{23}}$ $\mathbf{Al_2O_3}$ **formula units**.

A ***mole*** *is the amount of a substance that contains the same number of particles as there are atoms in 12.00 g of carbon-12.*

'Amount' may be **mass**, or **volume** if referring to a **gas**.

'Particles' may be **atoms**, **molecules** or **formula units**.

The mole and mass

In terms of mass, one mole is the **mass** of a substance that contains $\mathbf{6.02 \times 10^{23}}$ **particles** of that substance.

One mole of a substance has a mass equal to the relative atomic, molecular or formula mass expressed in **grams**.

Molar mass *is the relative atomic, molecular or formula mass expressed in grams.*

Examples

1 mole of **sodium, Na**: has a mass of **23 g**
and contains $\mathbf{6.02 \times 10^{23}}$ **Na atoms**.

1 mole of **carbon dioxide, CO_2**: has a mass of $12 + (2 \times 16)$ g = **44 g**
and contains $\mathbf{6.02 \times 10^{23}}$ $\mathbf{CO_2}$ **molecules**.

1 mole of **zinc oxide, ZnO**: has a mass of $65 + 16$ g = **81 g**
and contains $\mathbf{6.02 \times 10^{23}}$ **ZnO formula units**.

Some conversions

(a) Mass to number of moles:

$$\text{Number of moles} = \frac{\text{Mass}}{\text{Mass of one mole}}$$

(b) Number of moles to mass:

$$\text{Mass} = \text{Number of moles} \times \text{Mass of one mole}$$

(c) Number of particles to number of moles:

$$\text{Number of moles} = \frac{\text{Number of particles}}{6.02 \times 10^{23}}$$

(d) Number of moles to number of particles:

$$\text{Number of particles} = \text{Number of moles} \times 6.02 \times 10^{23}$$

i.e. $\quad$ Mass of element or compound $\xrightleftharpoons[\text{(b)}]{\text{(a)}}$ **Number of moles** $\xrightleftharpoons[\text{(d)}]{\text{(c)}}$ Number of particles

Sample questions

1 How many moles are there in 42 g of nitrogen?

Mass of 1 mole $N_2 = 2 \times 14$ g $= 28$ g

$\therefore \qquad$ Number of moles in **42 g N_2** $= \frac{42}{28}$ mole $\quad$ from equation (a)

$$= \textbf{1.5 mole}$$

2 What is the mass of 0.25 mole of aluminium sulphate?

Mass of 1 mole $Al_2(SO_4)_3 = (2 \times 27) + (3 \times 32) + (12 \times 16)$ g

$$= 342 \text{ g}$$

$\therefore \quad$ Mass of **0.25 mole $Al_2(SO_4)_3$** $= 0.25 \times 342$g $\quad$ from equation (b)

$$= \textbf{85.5 g}$$

3 How many moles of ammonia contain 18.06×10^{23} molecules of ammonia?

1 mole NH_3 contains 6.02×10^{23} NH_3 molecules.

$\therefore \qquad$ Number of moles in **18.06×10^{23} NH_3 molecules**

$$= \frac{18.06 \times 10^{23}}{6.02 \times 10^{23}} \text{ moles} \quad \text{from equation (c)}$$

$$= \textbf{3 moles}$$

4 How many sodium chloride formula units are there in 0.5 mole of sodium chloride?

1 mole NaCl contains 6.02×10^{23} NaCl formula units.

$\therefore \quad$ **0.5 mole NaCl** contains $0.5 \times 6.02 \times 10^{23}$ NaCl formula units

from equation (d)

$$= \textbf{3.01} \times \textbf{10}^{\textbf{23}} \textbf{ NaCl formula units}$$

5 How many hydrogen molecules are there in 10 g of hydrogen?

$$\text{Mass of 1 mole } H_2 = 2 \times 1 \text{ g} = 2 \text{ g}$$

$$\therefore \quad \text{Number of moles in } \mathbf{10 \text{ g } H_2} = \frac{10}{2} \text{ moles} \quad \text{from equation (a)}$$

$$= \mathbf{5 \text{ moles}}$$

$$\text{1 mole } H_2 \text{ contains } 6.02 \times 10^{23} \text{ } H_2 \text{ molecules.}$$

$$\therefore \quad \mathbf{5 \text{ moles } H_2} \text{ contain } 5 \times 6.02 \times 10^{23} \text{ } H_2 \text{ molecules} \quad \text{from equation (d)}$$

$$= \mathbf{3.01 \times 10^{24} \text{ } H_2 \text{ molecules}}$$

The mole and gas volumes

AVOGADRO'S LAW

Equal volumes of all gases at the same temperature and pressure contain the same number of molecules.

Since 1 mole of any gas contains 6.02×10^{23} molecules of that gas, if the number of molecules in each gas is 6.02×10^{23}, then it follows that **1 mole** of all gases under the same conditions of temperature and pressure have the **same volume**.

Molar volume *is the volume occupied by 1 mole of a gas.*

- At standard temperature and pressure – **s.t.p.** (0 °C, 760 mm Hg) – molar volume = **22.4 dm³**.
- At room temperature and pressure – **r.t.p.** (20 °C, 760 mm Hg) – molar volume = **24 dm³**.

Example

1 mole of **hydrogen, H_2:** has a volume of **22.4 dm³** at s.t.p.

$$\text{or } \mathbf{24 \text{ dm}^3} \text{ at r.t.p.}$$

$$\text{and contains } \mathbf{6.02 \times 10^{23} \text{ } H_2} \text{ molecules.}$$

Some conversions

(e) Volume of gas to number of moles:

$$\text{Number of moles} = \frac{\text{Volume of gas}}{22.4 \text{ dm}^3} \quad \text{(at s.t.p.)}$$

$$\text{Number of moles} = \frac{\text{Volume of gas}}{24 \text{ dm}^3} \quad \text{(at r.t.p.)}$$

(f) Number of moles to volume of gas:

$$\text{Volume of gas} = \text{Number of moles} \times 22.4 \text{ dm}^3 \quad \text{(at s.t.p.)}$$

$$\text{Volume of gas} = \text{Number of moles} \times 24 \text{ dm}^3 \quad \text{(at r.t.p.)}$$

i.e.

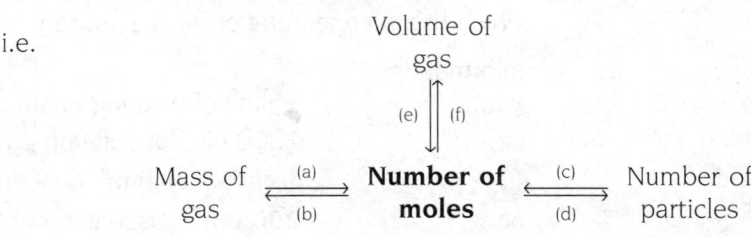

(For (a) to (d), see p. 48.)

Sample questions

1 How many moles are there in 1.12 dm^3 of ammonia at s.t.p.?

$$\text{Volume of 1 mole } NH_3 \text{ at s.t.p.} = 22.4 \ dm^3$$

$$\therefore \quad \text{Number of moles in } \mathbf{1.12 \ dm^3 \ NH_3} = \tfrac{1.12}{22.4} \text{ mole} \quad \text{from equation (e)}$$

$$= \mathbf{0.05 \ mole}$$

2 What volume is occupied by 0.1 mole of carbon dioxide at r.t.p.?

$$\text{Volume of 1 mole } CO_2 \text{ at r.t.p.} = 24 \ dm^3$$

$$\therefore \quad \text{Volume of } \mathbf{0.1 \ mole \ CO_2} = 0.1 \times 24 \ dm^3 \quad \text{from equation (f)}$$

$$= \mathbf{2.4 \ dm^3}$$

3 What is the volume of 17.75 g of chlorine at s.t.p.?

$$\text{Mass of 1 mole } Cl_2 = 2 \times 35.5 \ g = 71 \ g$$

$$\therefore \quad \text{Number of moles in } \mathbf{17.75 \ g \ Cl_2} = \tfrac{17.75}{71} \text{ mole from equation (a)}$$

$$= \mathbf{0.25 \ mole}$$

$$\text{Volume of 1 mole } Cl_2 \text{ at s.t.p.} = 22.4 \ dm^3$$

$$\therefore \quad \text{Volume of } \mathbf{0.25 \ mole \ Cl_2} = 0.25 \times 22.4 \ dm^3 \quad \text{from equation (f)}$$

$$= \mathbf{5.6 \ dm^3}$$

The mole and solutions

The **concentration** of a solution may be measured in **two** ways:

- **Grams per cubic decimetre** of solution, i.e. **g dm^{-3}**.
- **Moles per cubic decimetre** of solution, i.e. **mol dm^{-3}**.

Example

1 dm^3 of a solution of sodium hydroxide contains 20 g of sodium hydroxide:

$$\text{Mass of 1 mole } NaOH = 40 \ g$$

$$\therefore \quad \text{Number of moles in } \mathbf{20 \ g \ NaOH} = \tfrac{20}{40} \text{ mole}$$

$$= \mathbf{0.5 \ mole}$$

$$\therefore \quad \text{The solution has a concentration of } \mathbf{20 \ g \ dm^{-3}} \text{ or } \mathbf{0.5 \ mol \ dm^{-3}}.$$

A **standard solution** is one whose exact concentration is known. The solution is prepared using a **volumetric flask**.

Sample questions

1 How many moles of sodium sulphate are present in 200 cm^3 of a solution which has a concentration of 0.1 mol dm^{-3}?

Information
given: 1 dm^3 of solution contains 0.1 mole Na_2SO_4

i.e.   1000 cm^3 of solution contain 0.1 mole Na_2SO_4

$\therefore$ 1 cm^3 of solution contains $\tfrac{0.1}{1000}$ mole Na_2SO_4

and   **200 cm^3** of solution contain $200 \times \tfrac{0.1}{1000}$ mole Na_2SO_4

$$= \mathbf{0.02 \ mole \ Na_2SO_4}$$

2 What mass of sodium carbonate must be dissolved in 100 cm^3 of solution to produce a solution that has a concentration of 0.2 mol dm^3?

$\qquad$ 1000 cm^3 of the required solution contain 0.2 mole Na$_2$CO$_3$

∴ $\qquad$ 1 cm^3 of the required solution contains $\frac{0.2}{1000}$ mole Na$_2$CO$_3$

and

100 cm^3 of the required solution contain $100 \times \frac{0.2}{1000}$ mole Na$_2$CO$_3$

$\qquad\qquad\qquad\qquad\qquad$ = **0.02 mole Na$_2$CO$_3$**

$\qquad$ Mass of 1 mole Na$_2$CO$_3$ = $(2 \times 23) + 12 + (3 \times 16)$ g = 106 g

∴ $\qquad$ Mass of **0.02 mole Na$_2$CO$_3$** = 0.02 × 106 g

$\qquad\qquad\qquad\qquad\qquad$ = **2.12 g**

i.e. $\qquad$ Mass of Na$_2$CO$_3$ required = **2.12 g**

3 500 cm^3 of a solution of potassium hydroxide contain 2.8 g of potassium hydroxide. What is the concentration of this solution in mol dm^{-3}?

$\qquad$ 500 cm^3 of solution contain 2.8 g KOH

∴ $\qquad$ 1 cm^3 of solution contains $\frac{2.8}{500}$ g KOH

and $\qquad$ **1000 cm^3** of solution contain $1000 \times \frac{2.8}{500}$ g KOH

$\qquad\qquad\qquad\qquad\qquad$ = **5.6 g KOH**

$\qquad$ Mass of 1 mole KOH = 39 + 16 + 1 g = 56 g

∴ $\qquad$ Number of moles in **5.6 g KOH** = $\frac{5.6}{56}$ mole

$\qquad\qquad\qquad\qquad\qquad$ = **0.1 mole**

i.e. $\qquad$ Concentration of the solution = **0.1 mol dm^{-3}**

The mole and chemical formulae

The Law of Constant Composition *states that all pure samples of the same chemical compound contain the same elements combined together in the same proportions by mass.*

A **chemical formula** shows how many **moles** of each element combine to form **one mole** of the compound, e.g. H$_2$O represents **2 moles** of **hydrogen** atoms combined with **1 mole** of **oxygen** atoms.

- The **empirical formula** shows the **simplest mole ratio** of each element present in a compound.
- The **molecular formula** is a simple multiple of the empirical formula. It shows the **actual number** of moles of each element present in one mole of a **molecular** substance.

Both empirical and molecular formulae of compounds can be determined experimentally by finding the masses of each element in a given mass of the compound.

Example

A piece of magnesium ribbon was weighed and burnt in air. The magnesium oxide produced was then weighed:

$\qquad\qquad$ Mass of **magnesium** = **2.4 g**

$\qquad$ Mass of **magnesium** oxide produced = **4.0 g**

$\therefore$ Mass of **oxygen** present in the magnesium oxide $= 4.0 - 2.4$ g

$= \textbf{1.6 g}$

To determine the empirical formula of magnesium oxide, the **number of moles** of each element must be calculated:

	Mg	O
Mass of element	2.4 g	1.6 g
A_r of element	24	16
Number of moles of element	$\frac{2.4}{24}$	$\frac{1.6}{16}$
	$= 0.1$ mole	$= 0.1$ mole
Simplest mole ratio	1 :	1

Empirical formula $= \textbf{MgO}$

Sample questions

1 On analysis, a compound was found to contain 55.19% potassium, 14.62% phosphorus and 30.19% oxygen. Determine the empirical formula of this compound.

	K	P	O
Percentage of element	55.19%	14.62%	30.19%
Mass of element in 100 g of compound	55.19 g	14.62 g	30.19 g
A_r of element	39	31	16
Number of moles of element	$\frac{55.19}{39}$	$\frac{14.62}{31}$	$\frac{30.19}{16}$
	$= 1.415$	$= 0.472$	$= 1.887$
Simplest mole ratio	3 :	1 :	4

Empirical formula $= \textbf{K}_3\textbf{PO}_4$

2 3 g of carbon and 0.5 g of hydrogen combine with oxygen to make a compound whose mass is 7.5 g and whose relative molecular mass is 60. Determine the molecular formula of this compound.

	C	H	O
Mass of element	3 g	0.5 g	$7.5 - 3.5$ g $= 4$ g
A_r of element	12	1	16
Number of moles of element	$\frac{3}{12}$	$\frac{0.5}{1}$	$\frac{4}{16}$
	$= 0.25$	$= 0.5$	$= 0.25$
Simplest mole ratio	1 :	2 :	1

Empirical formula $= \textbf{CH}_2\textbf{O}$

To find the **molecular formula**:

Relative molecular mass of $\textbf{CH}_2\textbf{O} = 12 + (2 \times 1) + 16 = \textbf{30}$

Relative molecular mass of **compound** $= \textbf{60}$

And $\frac{60}{30} = 2$

$\therefore$ Molecular formula is **2** $\times$ empirical formula

Molecular formula $= \textbf{C}_2\textbf{H}_4\textbf{O}_2$

PERCENTAGE COMPOSITION FROM FORMULAE

If the formula of a compound is **known**, the percentage composition, by mass, of each element present can be calculated.

Sample question

What is the percentage, by mass, of nitrogen in ammonium nitrate?

$$\text{Formula: } NH_4NO_3$$

$$\text{Mass of 1 mole } NH_4NO_3 = 14 + (4 \times 1) + 14 + (3 \times 16) \text{ g} = 80 \text{ g}$$

$$\text{Mass of } \textbf{nitrogen} \text{ in 1 mole } NH_4NO_3 = 14 + 14 \text{ g} = 28 \text{ g}$$

$$\therefore \quad \% \text{ nitrogen} = \frac{28}{80} \times 100$$

$$= \textbf{35\%}$$

The mole and chemical reactions

The Law of Conservation of Matter *states that matter can be neither created nor destroyed during a chemical reaction.*

In any chemical reaction, the total mass of the products is **equal** to the total mass of the original reactants.

Example

	$MgO(s)$	$+$	$2HCl(aq)$	$\longrightarrow$	$MgCl_2(aq)$	$+$	$H_2O(l)$

This means: 1 mole MgO   +   2 moles HCl $\longrightarrow$ 1 mole $MgCl_2$ +   1 mole H_2O

i.e. $(24 + 16)$ g MgO $+$ $2(1 + 35.5)$ g HCl $\longrightarrow$ $(24 + 71)$ g $MgCl_2$ $+$ $(2 + 16)$ g H_2O

     40 g MgO   +   73 g HCl $\longrightarrow$ 95 g $MgCl_2$ +   18 g H_2O

$$\underbrace{\qquad\qquad} \qquad\qquad \underbrace{\qquad\qquad}$$

         113 g $\longrightarrow$ 113 g

Points to note when answering questions

1. Begin by writing a **balanced equation** for the reaction. If the question involves **ions** then the **ionic equation** must be written.
2. Decide which reactant(s) and/or product(s) the question is **concerned with** and only deal with these.
3. Remember quantities of **gaseous** reactants and products may be measured in **mass** or **volume**.

Sample questions

1. What mass of sodium hydroxide must be added to sulphuric acid to produce 7.1 g of sodium sulphate?

     Equation: $2NaOH(s) + H_2SO_4(aq) \longrightarrow Na_2SO_4(aq) + 2H_2O(l)$

     Concerned with :   **2 moles NaOH** form **1 mole Na_2SO_4**

     Number of moles of Na_2SO_4 produced:

$$\text{Mass of 1 mole } Na_2SO_4 = (2 \times 23) + 32 + (4 \times 16) \text{ g} = 142 \text{ g}$$

$$\therefore \quad \text{Number of moles in 7.1 g} = \frac{7.1}{142} \text{ mole}$$

$$= \textbf{0.05 mole}$$

Since 2 moles NaOH reacts with 1 mole Na_2SO_4

2×0.05 mole NaOH reacts with 0.05 moles Na_2SO_4

i.e. **0.1 mole NaOH** reacts with **0.05 mole Na_2SO_4**

Mass of 1 mole NaOH $= 23 + 16 + 1$ g $= 40$ g

$\therefore$ Mass of 0.1 mole NaOH $= 0.1 \times 40$ g

$= \textbf{4.0 g}$

i.e. Mass of sodium hydroxide which must be used is **4.0 g**.

2 What volume of hydrogen, measured at s.t.p., will be produced when 8.1 g of aluminium react with hydrochloric acid?

Equation: $2Al(s) + 6HCl(aq) \longrightarrow 2AlCl_3(aq) + 3H_2(g)$

Concerned with: **2 moles Al** form **3 moles H_2**

Number of moles of Al used:

Mass of 1 mole Al $= 27$ g

$\therefore$ Number of moles in 8.1 g $= \frac{8.1}{27}$ moles

$= \textbf{0.3 mole}$

Since 2 moles Al form 3 moles H_2

1 mole Al forms $\frac{3}{2}$ moles H_2

and 0.3 mole Al forms $0.3 \times \frac{3}{2}$ mole H_2

i.e. **0.3 mole Al** forms **0.45 mole H_2**

Volume of 1 mole H_2 at s.t.p. $= 22.4$ dm^3

$\therefore$ Volume of 0.45 mole H_2 at s.t.p. $= 0.45 \times 22.4$ dm^3

$= \textbf{10.08 dm}^3$

i.e. Volume of hydrogen produced is **10.08 dm^3**.

3 What is the minimum mass of zinc carbonate that must be added to 250 cm^3 of 0.2 mol dm^{-3} hydrochloric acid to exactly neutralise the acid?

Equation: $ZnCO_3(s) + 2HCl(aq) \longrightarrow ZnCl_2(aq) + H_2O(l) + CO_2(g)$

Concerned with: **1 mole $ZnCO_3$** reacts with **2 moles HCl**

Number of moles of HCl used:

1000 cm^3 HCl(aq) contain 0.2 mole HCl

$\therefore$ 250 cm^3 HCl(aq) contain $250 \times \frac{0.2}{1000}$ mole HCl

$= \textbf{0.05 mole HCl}$

Since 1 mole $ZnCO_3$ reacts with 2 moles HCl

$\frac{1}{2}$ mole $ZnCO_3$ reacts with 1 mole HCl

and $0.05 \times \frac{1}{2}$ mole $ZnCO_3$ reacts with 0.05 mole HCl

i.e. **0.025 mole $ZnCO_3$** reacts with **0.05 mole HCl**

Mass of 1 mole $ZnCO_3 = 65 + 12 + 48$ g $= 125$ g

$\therefore$ Mass of **0.025 mole $ZnCO_3$** $= 0.025 \times 125$ g

$= \textbf{3.125 g}$

i.e. Minimum mass of zinc carbonate required is **3.125 g**.

4 What mass of lead (II) chloride would precipitate if a solution containing 8.28 g of lead(II) ions reacts with excess hydrochloric acid?

Ionic equation: $Pb^{2+}(aq) + 2Cl^-(aq) \longrightarrow PbCl_2(s)$

Concerned with: **1 mole Pb^{2+} ions forms 1 mole $PbCl_2$**

Number of moles of Pb^{2+} ions used:

Mass of 1 mole Pb^{2+} ions $= 207$ g

∴ Number of moles in 8.28 g $= \frac{8.28}{207}$ mole

$= $ **0.04 mole**

Since 1 mole Pb^{2+} ions forms 1 mole $PbCl_2$

0.04 mole Pb^{2+} ions forms **0.04 mole $PbCl_2$**

Mass of 1 mole $PbCl_2 = 207 + (2 \times 35.5)$ g $= 278$ g

∴ Mass of 0.04 mole $PbCl_2 = 0.04 \times 278$ g

$= $ **11.12 g**

i.e. Mass of lead(II) chloride produced is **11.12 g**

Quantitative analysis calculations

Quantitative (volumetric) analysis involves performing a series of **titrations** to determine the **exact volume** of one solution required to react completely with a **fixed volume** of another solution.

Titrations may involve:

- An **acid** and an **alkali** or **carbonate** – the **end point** (point at which the reaction is complete) may be determined using an **indicator** or **temperature change**.
- An **oxidising agent** and a **reducing agent** – the **end-point** is determined by a sudden **colour change**.

Results obtained from titrations may be used in **four** ways:

1 To make a **potassium**, **sodium** or **ammonium salt** (see Table 9.5, p. 44).
2 To calculate the **mole ratio** in which the reactants combine and hence write **equations** for titration reactions.
3 To calculate the **concentration** of one of the reactants used in a titration.
4 To determine the **basicity** of an **acid** used in a titration.

USING A TITRATION TO CALCULATE THE MOLE RATIO IN WHICH REACTANTS COMBINE

Sample calculation

25 cm^3 of sodium hydroxide solution of concentration 0.8 mol dm^{-3} neutralise 10 cm^3 of the acid H_2A containing 1.0 mol dm^{-3}. Calculate the mole ratio in which the reactants combine and hence write an equation for the reaction.

Method 1: Using first principles

Number of moles of NaOH used:

$$1000 \ cm^3 \ NaOH(aq) \ contain \ 0.8 \ mole \ NaOH$$

∴   **25 cm^3** NaOH(aq) contain $25 \times \frac{0.8}{1000}$ mole NaOH

$$= \textbf{0.02 mole NaOH}$$

Number of moles of H_2A used:

$$1000 \ cm^3 \ H_2A(aq) \ contain \ 1.0 \ mole \ H_2A$$

∴   **10 cm^3** H_2A(aq) contain $10 \times \frac{1.0}{1000}$ mole H_2A

$$= \textbf{0.01 mole } \textbf{H}_2\textbf{A}$$

i.e.   0.02 mole   NaOH reacts with 0.01 mole H_2A

∴   **2 moles**   **NaOH** react with **1 mole H_2A**.

Equation : $2NaOH(aq) + H_2A(aq) \longrightarrow Na_2A(aq) + 2H_2O(l)$

Method 2: Using the formula

$$\frac{M_a \times V_a}{M_b \times V_b} = \frac{\text{Number of moles of acid reacting}}{\text{Number of moles of alkali or carbonate reacting}}$$

M_a = Concentration of acid in **mol dm^{-3}**

V_a = Volume of acid in **cm^3**

M_b = Concentration of alkali or carbonate solution in **mol dm^{-3}**

V_b = Volume of alkali or carbonate solution in **cm^3**

Question states:

$$M_a = 1.0 \text{ mol dm}^{-3} \qquad M_b = 0.8 \text{ mol dm}^{-3}$$

$$V_a = 10 \text{ cm}^3 \qquad V_b = 25 \text{ cm}^3$$

$\therefore$

$$\frac{1.0 \times 10}{0.8 \times 25} = \frac{\text{Number of moles of H}_2\text{A reacting}}{\text{Number of moles of NaOH reacting}}$$

i.e.

$$\frac{\text{Number of moles of H}_2\text{A}}{\text{Number of moles of NaOH}} = \frac{10}{20}$$

$$= \frac{1}{2}$$

i.e.   **2 moles NaOH** react with **1 mole H$_2$A**.

Equation : $2NaOH(aq) + H_2A(aq) \longrightarrow Na_2A(aq) + 2H_2O(l)$

USING A TITRATION TO CALCULATE THE CONCENTRATION OF ONE OF THE REACTANTS

Sample calculation

25 cm^3 of hydrochyloric acid are needed to neutralise 50 cm^3 of potassium carbonate solution of concentration 69 g dm^{-3}. What is the concentration of the acid in mol dm^{-3}?

Method 1: Using first principles

Equation for the reaction:

$$K_2CO_3(aq) + 2HCl(aq) \longrightarrow 2KCl(aq) + H_2O(l) + CO_2(g)$$

Number of moles of K_2CO_3 used:

1000 cm^3 K$_2$CO$_3$(aq) contain 69 g K$_2$CO$_3$

$\therefore$ **50 cm^3** K$_2$CO$_3$(aq) contain $50 \times \frac{69}{1000}$ g K$_2$CO$_3$

$$= \textbf{3.45 g K}_2\textbf{CO}_3$$

Mass of 1 mole K$_2$CO$_3$ = 138 g

$\therefore$ Number of moles in **3.45 g K$_2$CO$_3$** $= \frac{3.45}{138}$ mole

$$= \textbf{0.025 mole K}_2\textbf{CO}_3$$

From the equation:

1 mole K$_2$CO$_3$ neutralises 2 moles HCl

$\therefore$ **0.025 mole K$_2$CO$_3$** neutralises 0.025×2 mole HCl

$$= \textbf{0.05 mole HCl}$$

Since 25 cm^3 HCl(aq) were used:

 25 cm^3 **HCl(aq)** must contain **0.05 mole HCl**

∴ 1000 cm^3 HCl(aq) contain $1000 \times \frac{0.05}{25}$ moles HCl

$$= \textbf{2.0 moles HCl}$$

i.e. Concentration of HCl(aq) is **2.0 mol dm^{-3}**.

Method 2: Using the formula

Equation for the reaction:

$$K_2CO_3(aq) + 2HCl(aq) \longrightarrow 2KCl(aq) + H_2O(l) + CO_2(g)$$

$$M_a = ? M_b = 69 \text{ g dm}^{-3}$$
$$V_a = 25 \text{ cm}^3 V_b = 50 \text{ cm}^3$$

Number of moles of HCl = 2 Number of moles of K$_2$CO$_3$ = 1

Concentration of K$_2$CO$_3$(aq) in mol dm^{-3}:

$$\text{Mass of 1 mole K}_2CO_3 = 138 \text{ g}$$

∴ Number of moles in **69 g** K$_2$CO$_3$ = $\frac{69}{138}$ mole

$$= \textbf{0.5 mole}$$

i.e. Concentration of K$_2$CO$_3$(aq) = **0.5 mol dm^{-3}**

Applying the formula:

$$\frac{M_a \times 25}{0.5 \times 50} = \frac{2}{1}$$

∴ $ M_a = \dfrac{2 \times 0.5 \times 50}{1 \times 25}$ mol dm^{-3}

$$= \textbf{2.0 mol dm}^{-3}$$

i.e. Concentration of HCl(aq) is **2.0 mol dm^{-3}**.

USING A TITRATION TO DETERMINE THE BASICITY OF AN ACID

Sample calculation

50 cm^3 of sodium hydroxide solution of concentration 0.5 mol dm^{-3} react with 25 cm^3 of an acid with the formula H$_n$A and concentration 0.5 mol dm^{-3}. Determine the value of n.

Using first principles

Number of moles of NaOH used:

 1000 cm^3 NaOH(aq) contain 0.5 mole NOH

∴ **50 cm^3** NaOH(aq) contain $50 \times \frac{0.5}{1000}$ mole NaOH

$$= \textbf{0.025 mole NaOH}$$

Number of moles of H$_n$A used:

 1000 cm^3 H$_n$A contain 0.5 mole H$_n$A

∴ **25 cm^3** H$_n$A contain $25 \times \frac{0.5}{1000}$ mole H$_n$A

$$= \textbf{0.0125 mole H}_n\textbf{A}$$

i.e.   0.025 mole NaOH reacts with 0.0125 mole H_nA

∴   **2 moles NaOH**   react with **1 mole H_nA**

Ionically: **2 moles OH$^-$ ions** react with **n moles H$^+$ ions**   [1]

Ionic equation for **any alkali** reacting with **any acid**:

$$OH^-(aq) + H^+(aq) \longrightarrow H_2O(l)$$

i.e.   1 mole OH$^-$ ions reacts with 1 mole H$^+$ ions

or   **2 moles OH$^-$ ions** react with **2 moles H$^+$ ions**   [2]

Comparing statements (1) and (2): **n = 2**

The acid is **dibasic**, with the formula H_2A.

12 | Types of chemical reactions

1 **Combination or synthesis** occurs when two or more substances chemical combine to form a single substance:

$$A \ + \ B \longrightarrow \ AB$$

e.g. $\quad 2FeCl_2(s) + Cl_2(g) \longrightarrow 2FeCl_3(s)$

2 **Decomposition** occurs when a compound splits up into simpler substances, usually owing to the action of heat (see page 92).

$$AB \longrightarrow A \ + \ B$$

e.g. $\quad CaCO_3(s) \xrightarrow{heat} CaO(s) + CO_2(g)$

3 **Displacement** occurs when an element in its **free state** takes the place of another element in a compound. Displacement reactions may be divided into **two** types:

- A **metal** may displace the **ions** of another **metal** or the **hydrogen ions** of an acid:

$$A \ + \ BX \longrightarrow \ AX \ + \ B$$

e.g. $\quad Mg(s) + CuSO_4(aq) \longrightarrow MgSO_4(aq) + Cu(s)$

$\quad Zn(s) + \ 2HCl(aq) \longrightarrow ZnCl_2(aq) \ + H_2(g)$

- A **non-metal** may displace the **ions** of another **non-metal**:

$$A \ + \ YB \longrightarrow \ YA \ + \ B$$

e.g. $\quad Cl_2(g) + 2KI(aq) \longrightarrow 2KCl(aq) + I_2(aq)$

Displacement reactions only occur if the element in its free state is **more reactive** than the element in the compound.

4 **Double decomposition** occurs when two compounds exchange **radicals**. Usually one of the products is insoluble and forms a **precipitate**:

$$AX \ + \ BY \longrightarrow \ AY \ + \ BX$$

e.g. $\quad AgNO_3(aq) + NaCl(aq) \longrightarrow AgCl(s) + NaNO_3(aq)$

5 **Neutralisation** occurs when an acid and a base react producing a salt and water (see p. 38).

6 **Redox** occurs when one reactant is **reduced** and the other is **oxidised** (see p. 62).

7 **Reversible reaction** occurs when the direction of a chemical change can be easily **reversed** by changing the **conditions** under which the reaction takes place:

$$A + B \rightleftharpoons C + D$$

e.g. When ammonium chloride is heated it decomposes into ammonia and hydrogen chloride:

$$NH_4Cl(s) \longrightarrow NH_3(g) + HCl(g)$$

If ammonia and hydrogen chloride mix at room temperature they react forming ammonium chloride:

$$NH_3(g) + HCl(g) \longrightarrow NH_4Cl(s)$$

The reaction is, therefore, **reversible** and may be written:

$$NHCl(s) \rightleftharpoons NH_3(g) + HCl(g)$$

It proceeds to the right on heating and to the left on cooling.

In a reversible reaction, if the products are not separated from each other or from the reaction, then it is possible for the reaction to proceed in **both directions** at the same time such that the system always contains a **mixture** of reactants and products.

Dynamic equilibrium is reached when the rate of the forward reaction equals the rate of the backward reaction. Dynamic equilibrium can only be established in a **sealed system** where no substances can enter or leave.

13 | *Oxidation and reduction*

Oxidation and reduction are opposite processes that occur together. A reaction where both reduction and oxidation are taking place is called a redox reaction. Oxidation and reduction may be defined in four ways – see Table 13.2, p. 64.

Oxidation number

An oxidation number indicates the oxidation state of an element in its free state or in a compound. Sometimes the oxidation number appears in the name of the compound where it refers to the element after whose name it is written, e.g. copper(II) oxide – oxidation number of copper is 2. An oxidation number can be assigned to each element in its free state or within a compound.

RULES FOR ASSIGNING OXIDATION NUMBERS

1 The oxidation number of all atoms of an element in its free state is zero,

    e.g. $Cu = 0$, N atoms in $N_2 = 0$

2 The oxidation number of elements that exist as simple ions in ionic compounds is the same as the charge on the ion,

    e.g.   in $AlCl_3$: $Al^{3+} = +3$, $Cl^- = -1$

3 The oxidation number of elements in radicals or in covalent compounds may vary; often they appear in the name of the ion or compound,

    e.g.   manganate(VII) ion, MnO_4^-
                     – oxidation number of manganese $= +7$

                 nitrogen(IV) oxide, NO_2
                     – oxidation number of nitrogen $= +4$

    See Table 13.1.

4 The sum of all oxidation numbers of elements in a compound is zero,

    e.g. Al_2O_3:

        2(oxidation number of Al) + 3(oxidation number of O) = 0

              2(+3)   +   3(−2)   = 0
              +6   +   −6   = 0

5 The sum of all oxidation numbers of elements in a radical is equal to the charge on the ion,

e.g. SO_4^{2-}:

$$(\text{oxidation number of S}) + 4(\text{oxidation number of O}) = -2$$
$$(\text{oxidation number of S}) + \qquad 4(-2) \qquad = -2$$
$$(\text{oxidation number of S}) + \qquad -8 \qquad = -2$$
$$\text{oxidation number of S} = -2 + 8$$
$$= +\mathbf{6}$$

The SO_4^{2-} ion may, therefore, be called the **sulphate(VI) ion**.

Element	Oxidation number	Notes
Hydrogen	**+1**	Except in **hydrides** of **metals**, e.g. -1 in NaH and CaH_2
Chlorine Bromine Iodine	**−1**	Except when present in **radicals**, e.g. $+1$ in ClO^-; $+3$ in ClO_2^-; $+5$ in ClO_3^-, BrO_3^- and IO_3
Oxygen	**−2**	Except in **peroxides**, e.g. -1 in H_2O_2 and Na_2O_2
Sulphur	**−2**	Except when present in **radicals** or **covalent compounds**, e.g. $+4$ in SO_3^{2-} and SO_2; $+6$ in HSO_4^-, SO_4^{2-} and SO_3
Nitrogen	**−3**	Except when present in **radicals** or **covalent compounds**, e.g. -3 in NH_4^+ and NH_3; $+2$ in NO; $+3$ in NO_2^-; $+4$ in NO_2; $+5$ in NO_3^-
Carbon	**Varies**	e.g. -4 in CH_4; $+2$ in CO; $+4$ in HCO_3^-, CO_3^{2-} and CO_2
Transition metals when present in radicals, e.g. MnO_4^-, $Cr_2O_7^{2-}$	**Vary**	Oxidation number of metal appears in **name** of ion, e.g. $Cr_2O_7^{2-}$ is the dichromate(VI) ion – oxidation number of chromium is $+6$ (see sample question below)

Table 13.1 Rules for assigning oxidation numbers to elements whose oxidation numbers vary

Sample question

What is the oxidation number of chromium in the $Cr_2O_7^{2-}$ ion?

$$2(\text{oxidation number of Cr}) + 7(\text{oxidation number of O}) = -2$$
$$2(\text{oxidation number of Cr}) + \qquad 7(-2) \qquad = -2$$
$$2(\text{oxidation number of Cr}) + \qquad -14 \qquad = -2$$
$$2(\text{oxidation number of Cr}) = -2 + 14$$
$$= +12$$
$$\text{Oxidation number of Cr} = \frac{+12}{2}$$
$$= +\mathbf{6}$$

Definition in terms of	Oxidation	Reduction	Oxidising agent	Reducing agent	Example
Oxygen	**Gain** of oxygen	**Loss** of oxygen	**Gives** oxygen	**Removes** oxygen	$CuO(s) + H_2(g) \longrightarrow Cu(s) + H_2O(g)$ **H_2** has been **oxidised** – it has **gained** oxygen to form H_2O **CuO** has been **reduced** – it has **lost** oxygen to form Cu **CuO** is the **oxidising agent** – it **gave** oxygen to the H_2 **H_2** is the **reducing agent** – it **removed** oxygen from the CuO
Hydrogen	**Loss** of hydrogen	**Gain** of hydrogen	**Removes** hydrogen	**Gives** hydrogen	$H_2S(g) + Cl_2(g) \longrightarrow 2HCl(g) + S(s)$ **H_2S** has been **oxidised** – it has **lost** hydrogen to form S **Cl_2** has been **reduced** – it has **gained** hydrogen to form HCl **Cl_2** is the **oxidising agent** – it **removed** hydrogen from the H_2S **H_2S** is the **reducing agent** – it **gave** hydrogen to the Cl_2
Electrons	**Loss** of electrons (**OIL RIG** – Oxidation **Is** Loss, Reduction **Is** Gain)	**Gain** of electrons	**Removes** electrons	**Gives** electrons	$2Fe(s) + 3Cl_2(g) \longrightarrow 2FeCl_3(s)$ In forming FeCl$_3$, each **Fe** atom **loses** three electrons: $Fe - 3e^- \longrightarrow Fe^{3+}$ Overall: $2Fe(s) - 6e^- \longrightarrow 2Fe^{3+}(s)$ ∴ **Fe** has been **oxidised** – it has **lost** electrons to form the Fe^{3+} ion In forming FeCl$_3$, each **Cl** atom **gains** one electron: $Cl + e^- \longrightarrow Cl^-$ Overall: $3Cl_2(g) + 6e^- \longrightarrow 6Cl^-(s)$ ∴ **Cl_2** has been **reduced** – it has **gained** electrons to form Cl^- ions **Cl_2** is the **oxidising agent** – it **removed** electrons from the Fe **Fe** is the **reducing agent** – it **gave** electrons to the Cl_2
Oxidation number (see main text)	**Increase** in oxidation number of an element in its free state or an element within a compound	**Decrease** in oxidation number of an element in its free state or an element within a compound	**Causes an increase** in oxidation number of an element in its free state or an element within a compound	**Causes a decrease** in oxidation number of an element in its free state or an element within a compound	$2KI(aq) + Cl_2(g) \longrightarrow 2KCl(aq) + I_2(s)$ Oxidation numbers: +1, -1 0 +1, -1 0 **KI** has been **oxidised** – the oxidation number of the I^- ion has **increased** from **-1** to **0** **Cl_2** has been **reduced** – the oxidation number of the Cl atoms has **decreased** from **0** to **-1** **Cl_2** is the **oxidising agent** – it caused the oxidation number of the I^- ion to **increase** **KI** is the **reducing agent** – it caused the oxidation number of the Cl atoms to **decrease**

Table 13.2 Definitions and examples of oxidation and reduction

Recognising redox reactions

1 Write a **balanced equation** for the reaction.

2 Assign an **oxidation number** to each **different element** in its free state or combined within a compound. Write these numbers below the elements in the equation.

   Radicals which remain **unchanged** during a reaction may be **ignored**.

3 Determine the element whose oxidation number has **increased** – this has been **oxidised** by the other reactant. The other reactant is, therefore, the **oxidising agent**.

4 Determine the element whose oxidation number has **decreased** – this has been **reduced** by the other reactant. The other reactant is, therefore, the **reducing agent**.

5 If the oxidation number of **all** elements remains **unchanged** the reaction is **not** a redox reaction.

Example

$$Zn(s) + H_2SO_4(aq) \longrightarrow ZnSO_4(aq) + H_2(g)$$

Oxidation numbers: **0**   **+1**, –   **+2**, –   **0**

- The $SO_4{}^{2-}$ **ion** remains **unchanged**, therefore it may be disregarded.
- The oxidation number of **zinc** has **increased** from **0** to **+2**, therefore zinc has been **oxidised** by the sulphuric acid. Sulphuric acid is the oxidising agent.
- The oxidation number of the hydrogen in the sulphuric acid has **decreased** from **+1** to **0**, therefore the sulphuric acid has been **reduced** by the zinc. Zinc is the reducing agent.

Oxidising and reducing agents

TESTING FOR OXIDISING AND REDUCING AGENTS

When a substance acts as an **oxidising agent**, the substance itself is **reduced**. When a substance acts as a **reducing agent**, the substance itself is **oxidised**.

$$
\begin{array}{ccc}
\text{Oxidising} & & \text{Reducing} \\
\text{agent} & & \text{agent} \\
\mathbf{A} & + & \mathbf{B} \longrightarrow \mathbf{C + D} \\
\text{Reduced} & & \text{Oxidised} \\
\text{by } \mathbf{B} & & \text{by } \mathbf{A}
\end{array}
$$

- To test for an **oxidising agent**, mix the substance being tested with a **known** reducing agent that gives a **visible change** when it acts as a reducing agent, i.e. when it is **oxidised**.
- To test for a **reducing agent**, mix the substance being tested with a **known** oxidising agent that gives a **visible change** when it acts as an oxidising agent, i.e. when it is **reduced**.

Oxidising agent	Visible change when it acts as oxidising agent	Explanation of visible change
Acidified (H_2SO_4) potassium manganate(VII), **$KMnO_4/H^+$**	Purple to **colourless**	Purple MnO_4^- ion forms colourless Mn^{2+} ion
Acidified (H_2SO_4) potassium dichromate(VI), **$K_2Cr_2O_7/H^+$**	Orange to **green**	Orange $Cr_2O_7{}^{2-}$ ion forms green Cr^{3+} ion
Iron(III) salts, **Fe^{3+}**	Yellow to **pale green**	Yellow Fe^{3+} ion forms pale green Fe^{2+} ion
Dilute or concentrated nitric acid, **HNO_3**	**Brown** gas evolved	NO_2 produced – a brown gas
Concentrated sulphuric acid (usually hot), **H_2SO_4**	A **pungent** gas evolved	SO_2 produced – a pungent gas
Sodium chlorate(I) (bleach), **NaClO**	Depends on other reagent – turns many dyes **colourless**	Many dyes, when oxidised, lose their colour
Others: oxygen, **O_2**; chlorine, **Cl_2**; manganese(IV) oxide, **MnO_2**		

Table 13.3 Some common oxidising agents

Reducing agent	Visible change when it acts as reducing agent	Explanation of visible change
Potassium iodide, **KI**	Colourless to **brown**	$I_2(s)$ produced – dissolves in KI(aq) to form brown iodine solution
Hydrogen sulphide, **H_2S**	**Yellow** precipitate formed	$S(s)$ produced – yellow and insoluble
Iron(II) salts, **Fe^{2+}**	Pale green to **yellow**	Pale green Fe^{2+} ion forms yellow Fe^{3+} ion
Concentrated hydrochloric acid, **HCl**	**Yellow-green** gas evolved	Cl_2 produced – a yellow-green gas
Others: hydrogen, **H_2**; carbon, **C**; carbon monoxide, **CO**; reactive metals		

Table 13.4 Some common reducing agents

Oxidising agents		Reducing agents	
Test	Colour change if oxidising agent present	Test	Colour change if reducing agent present
Add **KI(aq)**	Colourless to **brown**	Add **KMnO$_4$(aq)** acidified with H$_2$SO$_4$(aq)	Purple to **colourless**
Bubble **H$_2$S(g)** through substance	**Yellow** precipitate produced	Add **K$_2$Cr$_2$O$_7$(aq)** acidified with H$_2$SO$_4$(aq)	Orange to **green**
Add aqueous **iron(II)** salt	Pale green to **yellow**	Add aqueous **iron(III)** salt	Yellow to **pale green**

Table 13.5 Summary of common tests for oxidising and reducing agents

SUBSTANCES THAT CAN BEHAVE AS BOTH OXIDISING AND REDUCING AGENTS

Sulphur dioxide (SO$_2$)

In the presence of **water**, sulphur dioxide is usually a **reducing agent**; if reacted with a **stronger** reducing agent, it acts as an **oxidising agent**:

- With **chlorine** it acts as a **reducing agent**:

$$SO_2(g) + 2H_2O(l) + Cl_2(g) \longrightarrow H_2SO_4(aq) + 2HCl(aq)$$

Oxidation numbers: +**4**,−2   +1, −2   **0**   +1,+**6**,−2   +1,−**1**

[reduction]

The oxidation number of the chlorine atoms in the chlorine has **decreased** from **0** to −**1**, therefore the chlorine has been **reduced** by the sulphur dioxide.

- With **hydrogen sulphide**, a stronger reducing agent, it acts as an **oxidising agent**:

$$2H_2S(g) + SO_2(g) \longrightarrow 3S(s) + 2H_2O(l)$$

Oxidation numbers: +1, −**2**   +**4**,−2   **0**   +1, −2

[oxidation]

The oxidation number of the sulphur in the hydrogen sulphide has **increased** from −**2** to **0**, therefore the hydrogen sulphide has been **oxidised** by the sulphur dioxide.

Acidified hydrogen peroxide (H$_2$O$_2$/H$^+$)

This is usually an **oxidising agent**; if reacted with a **stronger** oxidising agent, it acts as a **reducing agent**:

- With **potassium iodide** solution it acts as an **oxidising agent**, oxidising the iodide ions to iodine.
- With **acidified potassium manganate(VII)**, a stronger oxidising agent, it acts as a **reducing agent**, reducing the purple MnO$_4^-$ ion to the colourless Mn^{2+} ion.

14 Electrochemistry and electrolysis

Introduction

SOME DEFINITIONS

- **Electric current**: a flow of charged particles – electrons or ions (see below).
- **Conductor**: a substance that allows electricity to pass through, e.g. all metals, graphite (carbon).
- **Insulator (non-conductor)**: a substance that does not allow electricity to pass through, e.g. non-metals, plastics.
- **Electrolyte**: a molten substance or solution that allows electricity to pass through causing it to decompose, e.g. molten salts, aqueous solutions of acids, alkalis and salts.
- **Non-electrolyte**: a molten substance or solution that does not allow electricity to pass through, e.g. organic solvents, aqueous solutions of organic substances such as glucose, molten covalent substances such as wax.

METALLIC AND ELECTROLYTIC CONDUCTION

- **Metallic conduction** involves the flow of **free electrons** through a **metal**; these electrons are present in the electron pool. The metal remains chemically **unchanged**.
- **Electrolytic conduction** involves the movement of **free ions** through a molten substance or a solution. The substance is chemically **decomposed** during electrolytic conduction.

STRONG AND WEAK ELECTROLYTES

- **Strong electrolytes** are substances that are **fully ionised**, i.e. they contain a large number of ions, e.g. strong acids, strong alkalis, solutions of salts, molten salts.
- **Weak electrolytes** are substances that are **partially ionised**, i.e. they contain few ions, e.g. weak acids, weak alkalis, pure water.

A note about pure water

Pure water is an **extremely weak** electrolyte. Approximately one molecule in every 560 000 000 molecules is ionised:

$$H_2O(l) \rightleftharpoons H^+(aq) + OH^-(aq)$$

Electrolysis

Electrolysis is the **chemical change** occurring when an electric current is passed through an **electrolyte**.

Electrolysis involves the use of **electrodes** connected to a **battery**. The electrodes carry the current **into** or **out of** the electrolyte.

Electrodes are usually made of **inert (unreactive)** substances e.g. graphite (carbon), platinum:

- The **anode** is the **positive electrode**, connected to the **positive terminal** of the battery.
- The **cathode** is the **negative electrode**, connected to the **negative terminal** of the battery.

During electrolysis:

1 **Anions (−)** are attracted to the **anode (+)** where they **lose** electrons to form **neutral atoms**, i.e. they are **discharged**:

$$A^{n-} - ne^- \longrightarrow A$$

Oxidation therefore occurs at the **anode**.

2 **Electrons**, lost at the anode, are 'sucked along' to the positive terminal of the battery and pushed out of the negative terminal to the cathode.

3 **Cations (+)** are attracted to the **cathode (−)** where they **gain** the electrons from the anode to form **neutral atoms**, i.e. they are **discharged**:

$$C^{n+} + ne^- \longrightarrow C$$

Reduction therefore occurs at the **cathode**.

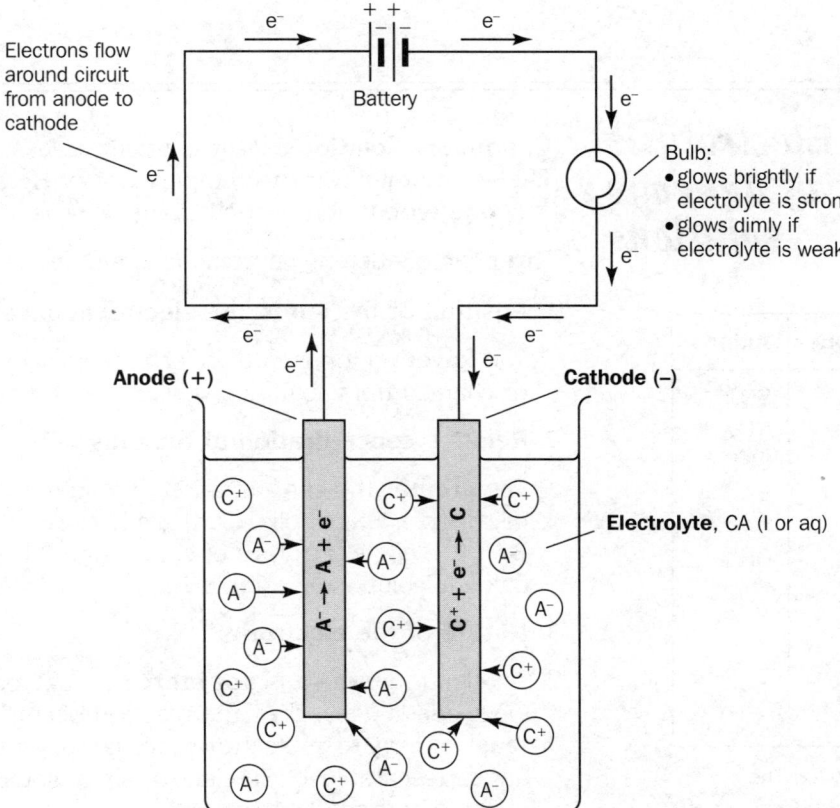

Fig. 14.1 The theory of electrolysis

Electrolysis of molten electrolytes

Molten electrolytes contain only **two** kinds of ion. Both are discharged during electrolysis.

Example

Electrolysis of molten lead(II) bromide using inert electrodes

Ions present: $Pb^{2+}(l)$, $Br^-(l)$

At anode:
Br^- ions are attracted to the anode and discharged, i.e. they **lose** electrons to form **bromine atoms**:

$$Br^-(l) - e^- \longrightarrow Br$$

Pairs of bromine atoms immediately join to form **bromine molecules**:

$$Br + Br \longrightarrow Br_2(g)$$

Overall reaction: $\qquad 2Br^-(l) - 2e^- \longrightarrow Br_2(g)$

At the temperature required to melt lead(II) bromide, brown **bromine vapour** is released.

At cathode:
Pb^{2+} ions are attracted to the cathode and discharged, i.e. they **gain** electrons to form **lead atoms**:

$$Pb^{2+}(l) + 2e^- \longrightarrow Pb(l)$$

Molten lead drips off the cathode.

Electrolysis of aqueous solutions

An **aqueous solution** usually contains at least **two** different **cations** and **two** different **anions** owing to the presence of H^+ and OH^- ions from the **water**. Only **one type** of ion of each charge is usually discharged at **each** electrode.

Three factors determine which ions are discharged:

1 **Position of the ion in the electrochemical series**. (See Table 14.1.)

The **lower** an ion is in the electrochemical series, the **more likely** it is to be discharged from solution.

2 **Relative concentration of the ions**.

The **greater** the concentration of an ion, the **more likely** it is to be discharged. This applies particularly to solutions containing **halide ions (Cl^-, Br^-, I^-)**, e.g. electrolysis of concentrated hydrochloric acid and sodium chloride solution (see Table 14.2).

3 **Nature of the electrode**.

An electrode which is **not inert** can take part in a reaction. If an **active** electrode is used, then the reaction occurring is the one that requires **least energy**, e.g. electrolysis of copper sulphate solution using a copper anode (see Table 14.2); electrolysis of sodium chloride solution using a mercury cathode (see p. 115).

Cations	Anions
K^+	SO_4^{2-}
Ca^{2+}	NO_3^-
Na^+	Cl^-
Mg^{2+}	Br^-
Al^{3+}	I^-
Zn^{2+}	OH^-
Fe^{2+}	
Pb^{2+}	
H^+	
Cu^{2+}	
Ag^+	

Table 14.1 The electrochemical series

Electrolyte/ions present	Electrodes	Electrode reactions and products	Changes in electrolyte
Dilute H$_2$SO$_4$(aq) From H$_2$O: H$^+$(aq), OH$^-$(aq) From H$_2$SO$_4$: H$^+$(aq), SO$_4^{2-}$(aq)	**Inert** – platinum or carbon	• **Anode: OH$^-$** discharged, lower in electrochemical series than SO$_4^{2-}$ $$4OH^-(aq) - 4e^- \longrightarrow 2H_2O(l) + O_2(g)$$ **Oxygen** gas evolved • **Cathode:** 2H$^+$(aq) + 2e$^-$ $\longrightarrow$ H$_2$(g) **Hydrogen** gas evolved • **Volume of gases:** 1 volume O$_2$: 2 volumes H$_2$ (for every 4 moles e$^-$, 1 volume, O$_2$ and 2 volumes H$_2$ are produced)	Acid becomes more **concentrated**: H$^+$(aq) and OH$^-$(aq) are removed leaving H$^+$(aq) and SO$_4^{2-}$(aq)
Concentrated HCl(aq) From H$_2$O: H$^+$(aq), OH$^-$(aq) From HCl: H$^+$(aq), Cl$^-$(aq)	**Inert** Anode: carbon to resist attack by Cl$_2$ Cathode: platinum or carbon	• **Anode: Cl$^-$** discharged, higher concentration than OH$^-$ $$2Cl^-(aq) - 2e^- \longrightarrow Cl_2(g)$$ **Chlorine** gas evolved • **Cathode:** 2H$^+$(aq) + 2e$^-$ $\longrightarrow$ H$_2$(g) **Hydrogen** gas evolved • **Volume of gases:** 1 volume Cl$_2$: 1 volume H$_2$	Acid becomes more **dilute**: H$^+$(aq) and Cl$^-$(aq) are removed leaving H$^+$(aq) and OH$^-$(aq)
Very dilute NaCl(aq) From H$_2$O: H$^+$(aq), OH$^-$(aq) From NaCl: Na$^+$(aq), Cl$^-$(aq)	**Inert** – platinum or carbon	• **Anode: OH$^-$** discharged, lower in electrochemical series than Cl$^-$ $$4OH^-(aq) - 4e^- \longrightarrow 2H_2O(l) + O_2(g)$$ **Oxygen** gas evolved • **Cathode: H$^+$** discharged, lower in electrochemical series than Na$^+$ $$2H^+(aq) + 2e^- \longrightarrow H_2(g)$$ **Hydrogen** gas evolved • **Volume of gases:** 1 volume O$_2$: 2 volumes H$_2$	Solution becomes more **concentrated**: H$^+$(aq) and OH$^-$(aq) are removed, leaving Na$^+$(aq) and Cl$^-$(aq)

(continued)

Table 14.2 Electrolysis of some aqueous solutions

Electrolyte/ions present	Electrodes	Electrode reactions and products	Changes in electrolyte
Concentrated NaCl(aq) From H_2O: $H^+(aq)$, $OH^-(aq)$ From NaCl: $Na^+(aq)$, $Cl^-(aq)$	**Inert** Anode: carbon to resist attack by Cl_2 Cathode: platinum or carbon	• **Anode: Cl⁻** discharged, higher concentration than OH^- $$2Cl^-(aq) - 2e^- \longrightarrow Cl_2(g)$$ **Chlorine** gas evolved • **Cathode: H⁺** discharged, lower in electrochemical series than Na^+ $$2H^+(aq) + 2e^- \longrightarrow H_2(g)$$ **Hydrogen** gas evolved • **Volume of gases:** 1 volume Cl_2: 1 volume H_2	Solution becomes **alkaline:** $H^+(aq)$ and $Cl^-(aq)$ are removed leaving $Na^+(aq)$ and $OH^-(aq)$
CuSO₄(aq) From H_2O: $H^+(aq)$, $OH^-(aq)$ From $CuSO_4$: $Cu^{2+}(aq)$, $SO_4^{2-}(aq)$	**Inert** – platinum or carbon	• **Anode: OH⁻** discharged, lower in electrochemical series than SO_4^{2-} $$4OH^-(aq) - 4e^- \longrightarrow 2H_2O(l) + O_2(g)$$ **Oxygen** gas evolved • **Cathode: Cu²⁺** discharged, lower in electrochemical series than H^+ $$Cu^{2+}(aq) + 2e^- \longrightarrow Cu(s)$$ Pink-brown **copper** deposited, cathode gets **thicker**	Solution becomes **acidic:** $Cu^{2+}(aq)$ and $OH^-(aq)$ are removed leaving $H^+(aq)$ and $SO_4^{2-}(aq)$ Solution becomes **paler blue:** $Cu^{2+}(aq)$ removed
CuSO₄(aq) From H_2O: $H^+(aq)$, $OH^-(aq)$ From $CuSO_4$: $Cu^{2+}(aq)$, $SO_4^{2-}(aq)$	Anode: **copper (active)** Cathode: platinum, carbon or copper	• **Anode: Cu(s)** from electrode converted to **Cu²⁺**, requires less energy than discharge of OH^- $$Cu(s) - 2e^- \longrightarrow Cu^{2+}(aq)$$ Anode gets **thinner** • **Cathode: Cu²⁺** discharged, lower in electrochemical series than H^+ $$Cu^{2+}(aq) + 2e^- \longrightarrow Cu(s)$$ **Copper** deposited, cathode gets **thicker**	**No change:** copper is transferred from anode to cathode

Table 14.2 *Continued*

Uses of electrolysis

1 INDUSTRIAL EXTRACTION OF REACTIVE METALS

(See also p. 94.)

Example *Extraction of sodium in the Down's cell*

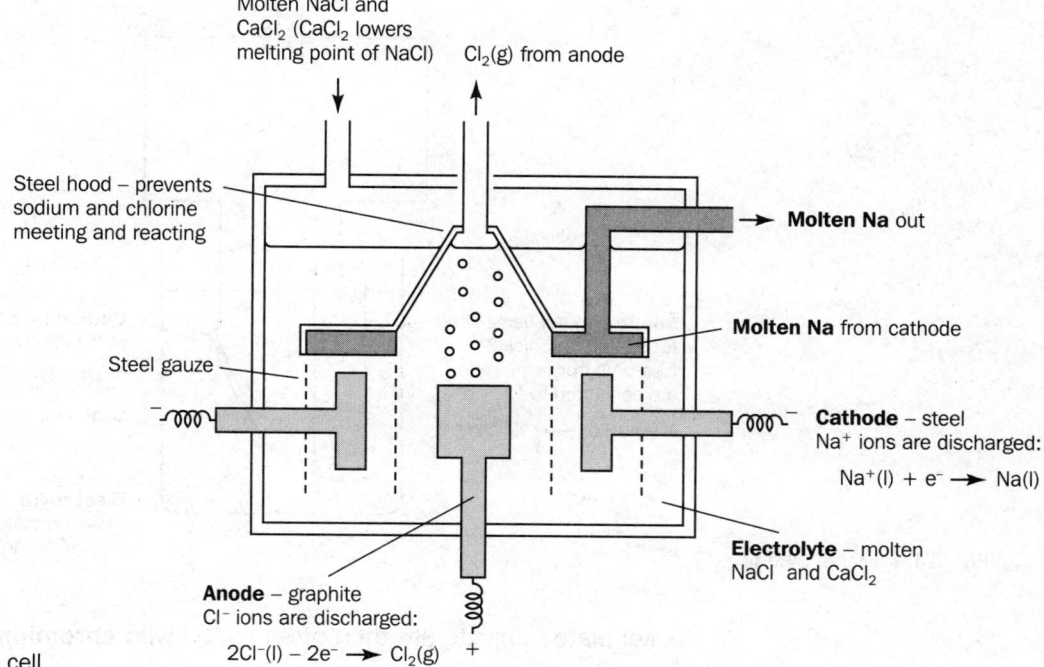

Molten NaCl and CaCl$_2$ (CaCl$_2$ lowers melting point of NaCl)

Cl$_2$(g) from anode

Steel hood – prevents sodium and chlorine meeting and reacting

Molten Na out

Molten Na from cathode

Steel gauze

Cathode – steel
Na$^+$ ions are discharged:

$$Na^+(l) + e^- \longrightarrow Na(l)$$

Electrolyte – molten NaCl and CaCl$_2$

Anode – graphite
Cl$^-$ ions are discharged:

$$2Cl^-(l) - 2e^- \longrightarrow Cl_2(g)$$ +

Fig. 14.2 The Down's cell

2 PURIFICATION OF METALS

Example *Purification of copper*

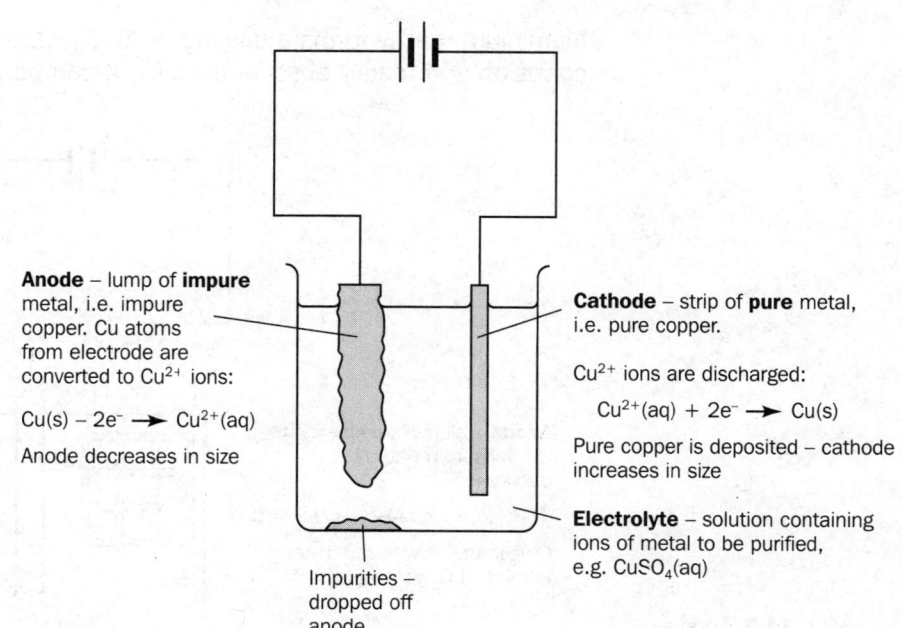

Anode – lump of **impure** metal, i.e. impure copper. Cu atoms from electrode are converted to Cu^{2+} ions:

$$Cu(s) - 2e^- \longrightarrow Cu^{2+}(aq)$$

Anode decreases in size

Cathode – strip of **pure** metal, i.e. pure copper.

Cu^{2+} ions are discharged:

$$Cu^{2+}(aq) + 2e^- \longrightarrow Cu(s)$$

Pure copper is deposited – cathode increases in size

Electrolyte – solution containing ions of metal to be purified, e.g. CuSO$_4$(aq)

Impurities – dropped off anode

Fig. 14.3 Purification of copper

3 ELECTROPLATING

Electroplating is the process of **coating** an object, usually metallic, with a **thin layer** of another **metal** by electrolysis. It is often used to prevent corrosion or to make an object more attractive.

Example   *Nickel plating*

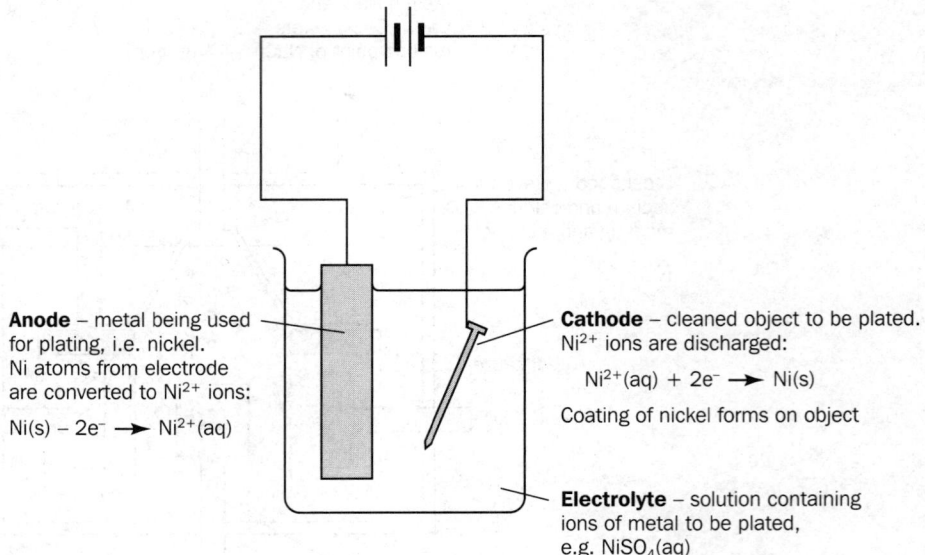

Anode – metal being used for plating, i.e. nickel. Ni atoms from electrode are converted to Ni^{2+} ions:

$$Ni(s) - 2e^- \longrightarrow Ni^{2+}(aq)$$

Cathode – cleaned object to be plated. Ni^{2+} ions are discharged:

$$Ni^{2+}(aq) + 2e^- \longrightarrow Ni(s)$$

Coating of nickel forms on object

Electrolyte – solution containing ions of metal to be plated, e.g. $NiSO_4(aq)$

Fig. 14.4 Nickel plating

Nickel-plated objects are then often plated with **chromium** or **silver**.

4 ANODISING

Anodising is a method of producing a **coating** on objects, such as window frames and saucepans, to make them **resistant to corrosion**.

Example   *Aluminium objects*

Aluminium readily forms a coating of Al_2O_3 which is coherent, resistant to corrosion and readily absorbs dyes (so it can be attractively coloured).

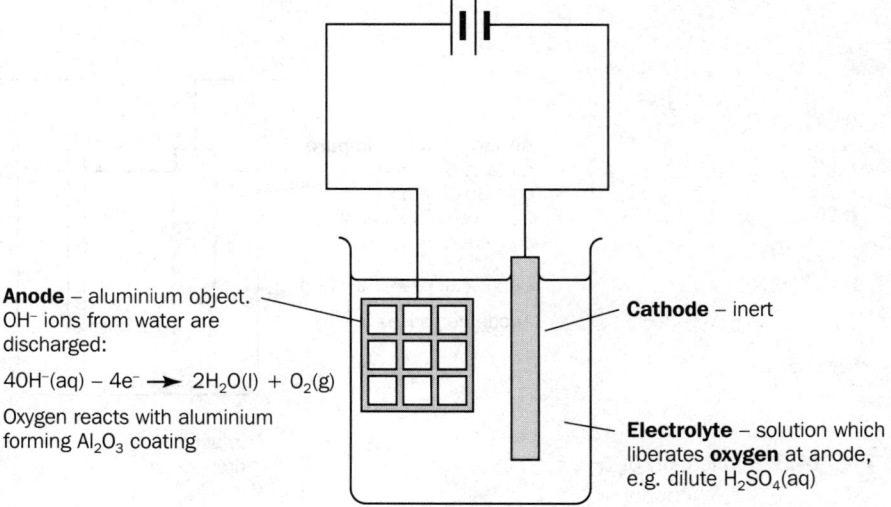

Anode – aluminium object. OH^- ions from water are discharged:

$$4OH^-(aq) - 4e^- \longrightarrow 2H_2O(l) + O_2(g)$$

Oxygen reacts with aluminium forming Al_2O_3 coating

Cathode – inert

Electrolyte – solution which liberates **oxygen** at anode, e.g. dilute $H_2SO_4(aq)$

Fig. 14.5 Anodising

5 OTHER USES

Other uses include the manufacture of chlorine and sodium hydroxide (see p. 115).

Quantitative electrolysis

Faraday's laws of electrolysis are concerned with the **quantity** of an element, measured in **moles**, formed during electrolysis.

1 FARADAY'S FIRST LAW OF ELECTROLYSIS

The mass of a substance produced at, or dissolved from, an electrode during electrolysis is directly proportional to the quantity of electricity passing through the electrolyte.

The **quantity of electricity** used depends on:

- The **length of time** a steady current is flowing through the electrolyte – measured in **seconds**.
- The **size** of that steady current – measured in **amperes (A or amps)**.

  i.e.   Quantity of electricity = Current × Time

Quantity of electricity is measured in **coulombs (C)** or **faradays**.

- **One coulomb** is **one ampere** flowing for **one second**.

  i.e.   Coulombs (C) = Amperes (A) × Seconds (s)

- **96 500 C = 1 faraday**

When **96 500 C** of electricity have flowed through a circuit, **one mole of electrons** or 6.02×10^{23} **electrons** have flowed.

Since **one mole of electrons** is required to discharge **one mole** of an ion with a **single charge**,

i.e. $\underset{\text{1 mole}}{A^-} - \underset{\text{1 mole}}{e^-} \longrightarrow \underset{\text{1 mole}}{A}$ or $\underset{\text{1 mole}}{C^+} + \underset{\text{1 mole}}{e^-} \longrightarrow \underset{\text{1 mole}}{C}$

it follows that **one faraday** is the quantity of electricity required to discharge **one mole** of an ion with a **single charge**.

96 500 C mol^{-1} is called the **Faraday constant**.

Example

During the electrolysis of molten sodium chloride, if 5 amperes flow for 32 minutes and 10 seconds, what mass of chlorine will be produced?

Time current flows = (32 × 60) + 10 seconds

= 1930 seconds

Quantity of electricity used = 5 × 1930 coulombs

= 9650 coulombs

$= \dfrac{9650}{96\,500}$ faraday

= 0.1 faraday

At anode: $2Cl^-(aq) - 2e^- \longrightarrow Cl_2(g)$

2 moles $\longrightarrow$ 1 mole

1 mole $\longrightarrow$ 0.5 mole

i.e. 1 mole Cl^- ions, when discharged, produce 0.5 mole Cl_2 and 1 faraday is required to discharge 1 mole of an ion with a **single** charge.

∴   **1 faraday** will discharge **1 mole Cl^- ions** producing **0.5 mole Cl_2**

i.e.   1 faraday produces 0.5 mole Cl_2

∴   **0.1 faraday** produces 0.1×0.5 mole Cl_2

$$= \textbf{0.5 mole } \textbf{Cl}_2$$

Mass of 1 mole $Cl_2 = 71$ g

∴   Mass of **0.05 mole Cl_2** $= 0.05 \times 71$ g

$$= \textbf{3.55 g}$$

i.e.   Mass of chlorine produced = **3.55 g**

2 FARADAY'S SECOND LAW OF ELECTROLYSIS

The number of faradays required to discharge one mole of an ion at an electrode is equal to the size of the charge on the ion.

- **1 faraday** is required to discharge one mole of an ion with a **single** charge, e.g. Na^+.
- **2 faradays** are required to discharge one mole of an ion with a **double** charge, e.g. Cu^{2+}.
- **3 faradays** are required to discharge one mole of an ion with a **triple** charge, e.g. Al^{3+}.

Example

During the electrolysis of copper(II) sulphate solution, if 0.5 faraday is used, what mass of copper will be deposited at the cathode?

At cathode: $Cu^{2+}(aq) + 2e^- \longrightarrow Cu(s)$

1 mole $\longrightarrow$ 1 mole

i.e. 1 mole Cu^{2+} ions, when discharged, produces 1 mole Cu and Cu^{2+} has a **double** charge.

∴   **2 faradays** will discharge **1 mole Cu^{2+} ions** producing **1 mole Cu**

i.e.   2 faradays produce 1 mole Cu

∴   1 faraday produces $\frac{1}{2}$ mole Cu

and   **0.5 faraday** produces $0.5 \times \frac{1}{2}$ mole Cu

$$= \textbf{0.25 mole Cu}$$

Mass of 1 mole Cu $= 64$ g

∴   Mass of **0.25 mole Cu** $= 0.25 \times 64$ g Cu

$$= \textbf{16 g Cu}$$

i.e.   Mass of copper deposited = **16 g**

Sample questions

1 Dilute sulphuric acid is electrolysed using platinum electrodes. If a current of 2.5 amperes flows for 25 minutes and 44 seconds, what mass and volume of oxygen, measured at s.t.p., will be formed at the anode?

$$\text{Time current flows} = (25 \times 60) + 44 \text{ seconds}$$
$$= \textbf{1544 seconds}$$
$$\text{Quantity of electricity used} = 2.5 \times 1544 \text{ coulombs}$$
$$= 3860 \text{ coulombs}$$
$$= \frac{3860}{96\,500} \text{ faraday}$$
$$= \textbf{0.04 faraday}$$

At anode: $4OH^-(aq) - 4e^- \longrightarrow 2H_2O(l) + O_2(g)$

4 moles $\longrightarrow$ 1 mole

1 mole $\longrightarrow$ 0.25 mole

i.e. 1 mole OH^- ions produces 0.25 mole O_2 and OH^- has a **single** charge.

∴   **1 faraday** discharges **1 mole OH^- ions** producing **0.25 mole O_2**

i.e. 1 faraday produces 0.25 mole O_2.

∴   **0.04 faraday** produces 0.04×0.25 mole O_2

$$= \textbf{0.01 mole } \mathbf{O_2}$$
$$\text{Mass of 1 mole } O_2 = 32 \text{ g}$$
$$\text{Mass of } O_2 \text{ produced} = 0.01 \times 32 \text{ g}$$
$$= \textbf{0.32 g}$$
$$\text{Volume of 1 mole } O_2 \text{ at s.t.p.} = 22.4 \text{ dm}^3$$
$$\text{Volume of } O_2 \text{ produced at s.t.p.} = 0.01 \times 22.4 \text{ dm}^3$$
$$= \textbf{0.224 dm}^3$$

2 What quantity of electricity is required to produce 5.4 g of aluminium during electrolysis of molten aluminium oxide?

At cathode: $Al^3(l) + 3e^- \longrightarrow Al(l)$

1 mole $\longrightarrow$ 1 mole

i.e. 1 mole Al^{3+} ions produces 1 mole Al and Al^{3+} has a **triple** charge.

∴   **3 faradays** discharge a **1 mole Al^{3+} ions** producing **1 mole Al**

i.e. 3 faradays produce 1 mole Al

Mass of 1 mole Al = 27 g

∴   **3 faradays** produce **27 g Al**

$\frac{3}{27}$ faraday produces 1 g Al

and $5.4 \times \frac{3}{27}$ faraday produces **5.4 g Al**

0.6 faraday produces 5.4 g Al

i.e. Quantity of electricity required is **0.6 faraday**.

77

15 Energy and chemical energetics

Forms of energy and their interconversions

Energy may be defined as the capacity to do work.

There are several **forms** of energy which are all interconvertible:

- **Kinetic**: energy resulting from the **movement** of an object
- **Potential**: energy due to the **position** of an object
- **Heat**
- **Electrical**
- **Chemical**: energy stored within a substance
- **Mechanical**
- **Light**
- **Sound**
- **Nuclear**.

The Law of Conservation of Energy states that energy can neither be created nor destroyed but merely changed from one form into another without any total loss or gain.

Examples of energy interconversions

1 Light bulb

nuclear energy $\longrightarrow$ **heat energy** $\longrightarrow$ **kinetic energy** $\longrightarrow$
stored in radioactive   as radioactive   as water boils
isotopes   nuclei split   forming steam

mechanical energy $\longrightarrow$ **electrical energy** $\longrightarrow$ **light energy**
as steam turns   as turbines generate   as light bulb
turbines   electricity   lights up

2 Whistling kettle

chemical energy $\longrightarrow$ **heat energy** $\longrightarrow$ **kinetic energy** $\longrightarrow$ **sound energy**
stored in   as gas burns   as water boils   as steam rushes
natural gas   forming steam   through whistle

Energy is measured in **joules (J)** or **kilojoules (kJ)**; 1000 J = 1 kJ.

Fuels

A **fuel** is a substance used to produce energy. Most fuels are compounds of **carbon** and **hydrogen** which liberate energy when **burnt** in air or oxygen. The energy, which is stored in fuels as **chemical energy**, was originally obtained from the Sun by **photosynthesis**.

Man's **major energy sources** are the **fossil fuels**, i.e. coal, crude oil, natural gas. Other fuels used by Man include wood, peat, charcoal and, of course, food.

When **choosing** a fuel, several factors must be considered:

- Cost
- Availability
- Ease of transportation
- Ease of storage
- Energy content
- Percentage of impurities
- Effect on the environment, e.g. when mined or burned.

Fuel	Origin	Mean energy value in kJ g^{-1}	Some disadvantages of use
Crude oil	Chemical alteration of small, ancient marine organisms by action of heat, pressure and anaerobic bacteria	45 (gasoline 43)	Danger of direct pollution by oil spillages during drilling and transporting. Burning produces air pollutants, e.g. CO_2, CO, SO_2 (see p. 111)
Natural gas	Same as for crude oil	55	Burning produces CO_2
Coal	Chemical alteration of wood from ancient forest by action of heat, pressure and anaerobic bacteria	25–33	Bulky to transport and store. Mining destroys environment. Burning produces CO_2, smoke, SO_2
Peat	First stage in coal formation before wood goes through heat and pressure changes	17	Bulky to transport and store. High water content that must be removed before burning. Burning produces CO_2, smoke, SO_2
Wood	Trees, grown as a result of photosynthesis	15	Bulky to transport and store. Deforestation. Burning produces CO_2, smoke
Charcoal ('coals')	Burning wood in a limited air supply	33	Bulky to transport and store. Deforestation. Burning produces CO_2

Table 15.1 Different fuels compared

Alternative energy sources

Reserves of fossil fuels are being used up at a rapid rate. These fuels are **non-renewable** and a major source of **pollution**, therefore if Man is to survive, he must develop **alternative energy sources** that are **renewable** and cause

little pollution. These energy sources include:

- Solar energy
- Wind energy
- Hydroelectricity
- Biogas
- Geothermal energy
- Gasohol
- Nuclear energy (see p. 15)
- Wave energy
- Tidal energy.

Examples

1 **Solar energy** – the energy of **sunlight**. Solar energy is used:

- To heat water in **solar water heaters**.
- To produce fresh water from salt water, or distilled water from tap water, in **solar stills**.
- To dry crops in **solar crop dryers**.
- To produce electrical energy in **solar cells**.

2 **Wind energy** – the energy resulting from the uneven heating of the Earth's surface. Wind energy is harnessed by **windmills** and used:

- To generate electricity.
- To pump water, e.g. for irrigation purposes.
- To circulate air for cooling purposes.

3 **Biogas** – a gas produced by **anaerobic fermentation** of plant and animal waste by bacteria in a **digester**. Biogas contains about 55–65% CH_4 (methane): 30–35% CO_2; 1% H_2S; and traces of H_2, CO and N_2. Biogas is used as a **fuel** for cooking, lighting, heating and refrigeration, especially in rural areas.

The **sludge**, left over after fermentation, is free of pathogens and odour, and is used as a **fertiliser**.

4 **Gasohol** – formed by mixing 20% ethanol and 80% gasoline. Gasohol is used as a fuel in motor vehicles. The ethanol is produced by **fermentation** of carbohydrates, e.g. molasses or cassava, using yeast (see p. 136).

Chemical energetics

All chemicals possess **energy** in **two** ways:

1 **Kinetic energy** as a result of the particles **vibrating** (solids) or **moving** (liquids and gases).

2 **Energy locked in chemical bonds**:

- When bonds are **broken**, energy is **absorbed**.
- When bonds are **formed**, energy is **released**.
- Making or breaking **strong** bonds (ionic, covalent, metallic) involve **large** energy changes.
- Making or breaking **weak** bonds (intermolecular forces) involve **small** energy changes.

ENERGY CHANGES DURING CHEMICAL REACTIONS

During a chemical reaction:

1 Bonds are **broken** in the **reactants** – energy is **absorbed**.
2 New bonds are **formed** in the **products** – energy is **released**.

Reactants	$\longrightarrow$	Products
Bonds **broken**		Bonds **formed**
Energy **absorbed**		Energy **released**

These energy changes occur **together** and can be measured as **changes** in **heat energy**:

1 The reaction is **endothermic** if:

$$\text{Energy } \textbf{absorbed} \text{ in breaking old bonds} > \text{Energy } \textbf{released} \text{ in forming new bonds}$$

Heat energy is **absorbed from** the surroundings and the reaction becomes **cooler** as this energy is absorbed, e.g. dissolving ammonium chloride, NH_4Cl, or potassium nitrate, KNO_3, in water.

2 The reaction is **exothermic** if:

$$\text{Energy } \textbf{absorbed} \text{ in breaking old bonds} < \text{Energy } \textbf{released} \text{ in forming new bonds}$$

Heat energy is **released to** the surroundings and the reaction becomes **hotter** as this energy is released, e.g. neutralisation reactions, dissolving ethanol in water, burning ethanol and other fuels.

ENTHALPY CHANGE OF A REACTION

The energy or heat content of any substance is called its **enthalpy**.

- **Enthalpy** is given the symbol **H**.
- **Change in enthalpy** is given the symbol ΔH

$$\text{Enthalpy change of a reaction} = \text{Energy content of } \textbf{products} - \text{Energy content of } \textbf{reactants}$$

i.e.
$$\Delta H_{\text{reaction}} = H_{\text{products}} - H_{\text{reactants}}$$

1 The reaction is **endothermic** if:

$$H_{\text{products}} > H_{\text{reactants}}$$

Heat energy is **absorbed** from the surroundings, ΔH is **positive** ($\Delta H + \textbf{ve}$).

2 The reaction is **exothermic** if:

$$H_{\text{products}} < H_{\text{reactants}}$$

Heat energy is **released** to the surroundings, ΔH is **negative** ($\Delta H - \textbf{ve}$).

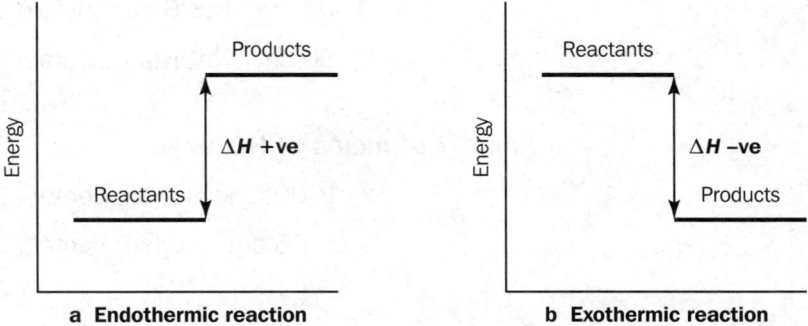

Fig. 15.1 Energy profile diagrams

a **Endothermic reaction**   b **Exothermic reaction**

ACTIVATION ENERGY

Activation energy is the **energy barrier** of a reaction. It is the energy that **reactants** must be given, in excess of the energy they normally possess, in order to start forming products.

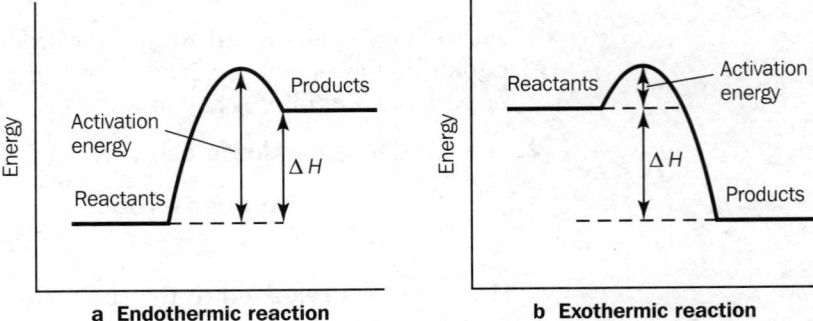

Fig. 15.2 Energy profile diagrams showing activation energy

HEAT OF NEUTRALISATION

Heat of neutralisation is the energy change which occurs when one mole of water is formed during a neutralisation reaction between an acid and an alkali.

The heat of neutralisation for **any** strong acid and **any** strong alkali is about -57 kJ mol^{-1}, since the energy change is for the common reaction:

$$OH^-(aq) + H^+(aq) \longrightarrow H_2O(l) \quad \Delta H = -57 \text{ kJ mol}^{-1}$$

$$\underset{\text{(J)}}{\overset{\text{Heat of}}{\underset{\text{neutralisation}}{}}} = \underset{\text{(g)}}{\overset{\text{Final mass}}{\underset{\text{of solution}}{}}} \times \underset{\text{(°C)}}{\overset{\text{Temperature}}{\underset{\text{rise}}{}}} \times \underset{\text{(J g}^{-1}\text{ °C}^{-1}\text{)}}{\overset{\text{Specific heat}}{\underset{\text{capacity of water}}{}}}$$

Specific heat capacity of water is **4.18 J g^{-1} °C^{-1}**. This is the amount of energy required to raise the temperature of 1 g of water by 1 °C.

Example

50 cm^3 of 1.0 mol dm^{-3} sodium hydroxide solution at room temperature was placed in a plastic cup. 25 cm^3 of 2.0 mol dm^{-3} hydrochloric acid, at the same temperature, was added to the alkali. The temperature of the mixture rose by 9.1 °C. Calculate the heat of neutralisation.

Number of moles of NaOH used:

$$1000 \text{ cm}^3 \text{ NaOH(aq) contain 1.0 mole NaOH}$$

$$\therefore \quad 50 \text{ cm}^3 \text{ NaOH(aq) contain } 50 \times \tfrac{1.0}{1000} \text{ mole NaOH}$$

$$= \textbf{0.05 mole NaOH}$$

Number of moles of HCl used:

$$1000 \text{ cm}^3 \text{ HCl(aq) contain 2.0 mole HCl}$$

$$\therefore \quad 25 \text{ cm}^3 \text{ HCl(aq) contain } 25 \times \tfrac{2.0}{1000} \text{ mole HCl}$$

$$= \textbf{0.05 mole HCl}$$

Equation: $\qquad NaOH(aq) + HCl(aq) \longrightarrow NaCl(aq) + H_2O(l)$

i.e. 1 mole NaOH reacts with 1 mole HCl forming 1 mole H_2O

$\therefore$ **0.05 mole NaOH** reacts with **0.05 mole HCl** forming **0.05 mole H_2O**

i.e. **0.05 mole** of water is formed.

$$\text{Final volume of solution} = 50 + 25 \text{ cm}^3$$
$$= 75 \text{ cm}^3$$

$$\text{Final mass of solution} = \textbf{75 g}$$

(assuming its density is the same as that of water: 1 g cm^{-3})

$$\text{Temperature rise} = \textbf{9.1 °C}$$

$\therefore$ $\quad$ Heat evolved in forming **0.05 mole H_2O** $= 75 \times 9.1 \times 4.18$ J

$$= \textbf{2852.85 J}$$

$\therefore$ $\quad$ Heat evolved in forming **1 mole H_2O** $= \dfrac{2852.85}{0.05}$ J

$$= 57\ 057 \text{ J}$$
$$= \textbf{57.057 kJ}$$

Since the reaction is **exothermic**, $\Delta H = \textbf{−57.06 kJ mol}^{-1}$

HEAT OF COMBUSTION

Heat of combustion *is the heat change which occurs when one mole of a substance in its normal state is completely burned in oxygen.*

Heat of combustion can be found by burning a fixed mass of the substance and allowing the heat produced to be transferred to a fixed mass of water. The temperature rise of the water is then recorded.

$$\begin{array}{c} \text{Heat of} \\ \text{combustion} \\ \text{(J)} \end{array} = \begin{array}{c} \text{Mass of} \\ \text{water heated} \\ \text{(g)} \end{array} \times \begin{array}{c} \text{Temperature} \\ \text{rise} \\ \text{(°C)} \end{array} \times \begin{array}{c} \text{Specific heat} \\ \text{capacity of water} \\ \text{(J g}^{-1}\text{ °C}^{-1}\text{)} \end{array}$$

Example

Burning 0.46 g of ethanol caused the temperature of 100 cm^3 of water to rise by 29.5 °C. Calculate the heat of combustion of ethanol.

$$\text{Mass of ethanol burned} = 0.46 \text{ g}$$

$$\text{Mass of 1 mole ethanol } (C_2H_5OH) = (24 + 5 + 16 + 1) \text{ g}$$
$$= 46 \text{ g}$$

$\therefore$ $\quad$ Number of moles in **0.46 g** $= \dfrac{0.46}{46}$ mole

$$= \textbf{0.01 mole}$$

i.e. **0.01 mole** ethanol is burned.

$$\text{Volume of water heated} = 100 \text{ cm}^3$$

$$\text{Mass of water heated} = \textbf{100 g}$$

$$\text{Temperature rise} = \textbf{29.5 °C}$$

$$\therefore \quad \text{Heat of combustion of } \textbf{0.01 mole} \text{ ethanol} = 100 \times 29.5 \times 4.18 \text{ J}$$
$$= \textbf{12 331 J}$$
$$\therefore \quad \text{Heat of combustion of } \textbf{1 mole} \text{ ethanol} = \frac{12\ 331}{0.01} \text{ J}$$
$$= 1\ 233\ 100 \text{ J}$$
$$= \textbf{1233.1 kJ}$$

Since the reaction is **exothermic**, $\Delta H = -\textbf{1233.1 kJ mol}^{-1}$

HEAT OF SOLUTION

Heat of solution is the energy change which occurs when one mole of a solute dissolves in such a volume of solvent that further dilution by the solvent produces no further heat change.

When **ionic** solids dissolve in water, **two** processes occur:

1 The ions are **pulled out** of the crystal lattice which involves breaking bonds – energy is **absorbed**.
2 Free ions become **associated** with water molecules, a process called **salvation** or **hydration** – energy is **released**.

The reaction is **endothermic** if **1 > 2**.
The reaction is **exothermic** if **2 > 1**.

$$\begin{array}{cccc} \text{Heat of} & \text{Mass of} & \text{Temperature} & \text{Specific heat} \\ \text{solution} = \text{solution} \times & \text{change} \times & \text{capacity of water} \\ \text{(J)} & \text{(g)} & (^\circ\text{C}) & (\text{J g}^{-1}\ {}^\circ\text{C}^{-1}) \end{array}$$

Example

Dissolving 10.1 g of potassium nitrate in 100 cm^3 of water resulted in a temperature decrease of 7.2 °C. Calculate the heat of solution of potassium nitrate.

$$\text{Mass of KNO}_3 \text{ dissolved} = 10.1 \text{ g}$$
$$\text{Mass of 1 mole KNO}_3 = 39 + 14 + (3 \times 16) \text{ g} = 101 \text{ g}$$
$$\therefore \quad \text{Number of moles in 10.1 g} = \frac{10.1}{101} \text{ mole}$$
$$= \textbf{0.1 mole}$$

i.e. **0.1 mole** KNO$_3$ is dissolved.

$$\text{Volume of water used} = 100 \text{ cm}^3$$
$$\text{Mass of solution formed} = \textbf{110.1 g}$$
$$\text{Temperature change} = \textbf{7.2 }^\circ\textbf{C}$$
$$\therefore \quad \text{Heat of solution of } \textbf{0.1 mole} \text{ KNO}_3 = 110.1 \times 7.2 \times 4.18 \text{ J}$$
$$= \textbf{3313.57 J}$$
$$\therefore \quad \text{Heat of solution of } \textbf{1 mole} \text{ KNO}_3 = \frac{3313.57}{0.1} \text{ J}$$
$$= 33\ 135.7 \text{ J}$$
$$= \textbf{33.14 kJ}$$

Since the reaction is **endothermic**, $\Delta H = +\textbf{33.14 kJ mol}^{-1}$

16 | Rates of reaction

The **rate of a reaction** is a measured change in the **amount** of a specific reactant or product per unit **time**:

$$\text{Rate} = \frac{\text{Measured change in a given property}}{\text{Time taken for change to occur}}$$

Some reactions occur very **rapidly**, e.g. precipitation, some occur very **slowly**, e.g. rusting of iron.

MEASURING RATES OF REACTION

During any reaction, the concentration of reactants **decreases** at the **same rate** as the concentration of products **increases**. Changes in concentration are not always easy to measure directly. **Property changes** which can be measured more easily include:

- **Volume** of a gaseous product
- **Colour intensity**
- **Amount** of a **precipitate** formed
- **Decrease in mass** (if a gas is evolved).

COLLISION THEORY FOR REACTIONS

In order to **react**:

1 Particles of the reacting substances must **collide** with each other to break old bonds in the reactants, and form new bonds in the products.
2 The colliding particles must produce **sufficient energy** to break the old bonds in the reactants. This energy is called the **activation energy** (see p. 82). Not all collisions produce the required energy, therefore not all collisions result in a reaction occurring.
3 Reactant particles must collide with the **correct orientation** so that the energy produced by the collision can be passed on to the bonds to be broken.

Factors affecting reaction rates

Reaction rates can be affected by:

- Concentration of reactants
- Pressure
- Temperature
- Particle size
- Catalysts
- Light.

See Table 16.1, p. 86.

Factor	Effect on reaction rate	Mechanism by which factor affects reaction rate	Examples of reactions that demonstrate effect
Concentration of reactants	If the concentration **increases**, the rate **increases**	Increasing concentration increases number of particles per unit volume – frequency of collisions increases	$Na_2S_2O_3(aq)$ (sodium thiosulphate) reacts with $HCl(aq)$ forming a precipitate of **sulphur**: $Na_2S_2O_3(aq) + 2HCl(aq) \longrightarrow$ $2NaCl(aq) + S(s) + H_2O(l) + SO_2(g)$ Repeat reaction varying concentration of $Na_2S_2O_3(aq)$ – measure time taken for a cross to disappear when viewed through the solution (i.e. fixed amount of precipitate obscures cross)
Pressure (affects gaseous reactions only)	If the pressure **increases**, the rate **increases**	Increasing pressure decreases volume – particles are closer together and therefore collide more frequently	—
Temperature	If the temperature **increases**, the rate **increases** – a **10 °C** rise **doubles** the rate	Increasing temperature increases kinetic energy of particles, causing them to move faster and: • collide more frequently, • collide with more energy so that more collisions produce sufficient energy to break bonds in the reactants	Iodine clock reaction – $KI(aq)$ reacts with acidified $H_2O_2(aq)$ to form **iodine** which turns blue in the presence of starch: $H_2O_2(aq) + 2KI(aq) + 2H^+(aq) \longrightarrow$ $I_2(aq) + 2K^+(aq) + 2H_2O(l)$ $Na_2S_2O_3$ delays the formation of the blue colour. Repeat reaction using same quantity of $Na_2S_2O_3(aq)$ each time, but at different temperatures – measure time taken for blue colour to appear
Particle size (of a solid reactant)	**Decreasing** the particle size (breaking the solid into smaller pieces) causes an **increase** in the rate	Reaction occurs at surface of solid – decreasing particle size increases surface area exposed, therefore gives more area for collisions	$CaCO_3(s) + 2HCl(aq) \longrightarrow CaCl_2(aq) + H_2O(l) + CO_2(g)$ Repeat above reaction using different sized lumps of $CaCO_3$, each with the same overall mass – measure rate of evolution of **CO_2**. Finely divided solids **catch fire** more readily than larger pieces of the same solid, e.g. in flour mills there is a danger of the flour burning explosively

Table 16.1 Factors affecting the rate of reactions

(continued)

Factor	Effect on reaction rate	Mechanism by which factor affects reaction rate	Examples of reactions that demonstrate effect
Catalyst	Adding a catalyst **increases** the rate (a negative catalyst or inhibitor decreases the rate)	Catalysts alter rate but remain chemically unchanged at the end. Catalysts provide an alternative pathway for a reaction that has a lower activation energy: 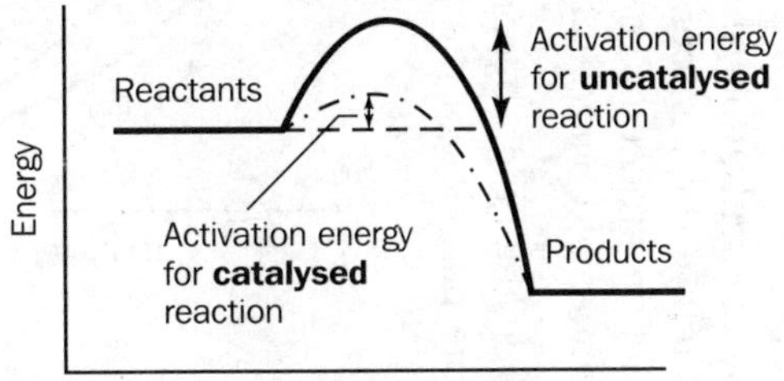 **Exothermic reaction**	H_2O_2 decomposes very slowly releasing O_2: $$2H_2O_2(aq) \longrightarrow 2H_2O(l) + O_2(g)$$ If **MnO$_2$** is added, decomposition is speeded up **Enzymes** are biological catalysts which speed up chemical reactions in living organisms **Tetraethyl lead(IV) (Pb(C$_2$H$_5$)$_4$)** is an inhibitor added to petrol to help stop premature ignition ('knocking')
Light	Light **increases** the rate of some reactions	Light is a form of energy which, on absorption, causes bonds to break in the reactants	$$H_2(g) + Cl_2(g) \longrightarrow 2HCl(g)$$ Above reaction does not occur in darkness but is explosive in bright light **Photosynthesis** only occurs in light

Table 16.1 *Continued*

Rate curves

When the **change in concentration** of a reactant or product is plotted against **time**, a **rate curve** is obtained.

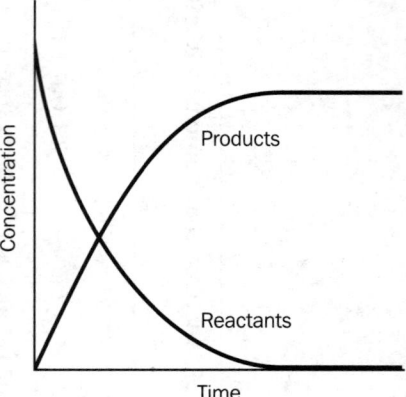

Fig. 16.1 Change in concentration of reactants and products during a reaction

The **rate** of any reaction is **greatest** where the graph is **steepest**, i.e. at the beginning of the reaction. The rate of any reaction **decreases** as the reaction proceeds.

The concentration of reactants is at its **highest** at the **start** of a reaction, therefore the number of collisions between particles is at its highest. As the reaction proceeds, the concentration of reactants **decreases**, therefore the number of collisions between particles decreases.

When the graph becomes **horizontal** the reaction is **complete**, i.e. there are no more particles of the reacting substances left to collide.

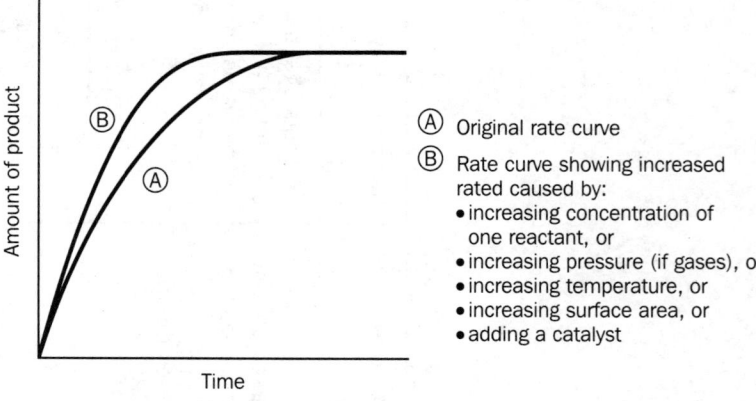

Ⓐ Original rate curve
Ⓑ Rate curve showing increased rated caused by:
• increasing concentration of one reactant, or
• increasing pressure (if gases), or
• increasing temperature, or
• increasing surface area, or
• adding a catalyst

Fig. 16.2 A rate curve showing the effect of any factor which increases the reaction rate

From Fig. 16.2:

1 Curve Ⓑ has a **steeper slope** – the rate of reaction is faster.
2 Curve Ⓑ becomes **horizontal sooner** – the reactants are used up faster.
3 Both curves become horizontal when the **same amount** of products has been produced – the original amount of reactants is not changed (when investigating concentration, it is the concentration of the reactant in excess which is changed; the amount of the limiting reactant is kept constant).

17 Metals and the reactivity series

A **metal** is an element which ionises by **losing** electrons, i.e. the element is **electropositive**.

General properties of metals

(See Table 2.2, p. 6 for physical properties.)

1 **Metals are reducing agents**.
   When a metal reacts it forms **positive ions** by **giving** electrons to the substance with which it is reacting:

   $$M - ne^- \longrightarrow M^{n+}$$

2 **Metals react with air or oxygen**.
   They form **basic oxides**, except aluminium, zinc and lead which form **amphoteric oxides**. If the oxide is soluble in water, it forms an alkaline solution, i.e. K_2O, Na_2O, CaO, (MgO is very slightly soluble).

3 **Most metals react with dilute acids**.
   With dilute hydrochloric and sulphuric acids, they form a **salt** and **hydrogen**.

4 **Most metals react with water or steam**.
   Those reacting with **water** form the **hydroxide** and **hydrogen**, those reacting with **steam** form the **oxide** and **hydrogen**.

5 **Metals generally react with non-metals**.
   They form **ionic** compounds.

The reactivity series of metals

The reactivity series of metals is a list of metals arranged in order of their reactivity based on:

- **Displacement** reactions (see p. 90).
- Reactions with **oxygen**, **dilute acids** and **water** (see Table 17.2, p. 91).
- **Reduction** of their **oxides** by hydrogen or carbon (see Table 17.2, p. 91).
- Relative ease of **decomposition** of their **compounds**, i.e. nitrates, carbonates and hydroxides (see Table 17.3, p. 92).

Metals at the **top** of the series are the **most electropositive**, i.e. they **lose** electrons very **easily** and therefore readily form ions. Their ions are very **stable** and are not easily discharged to form atoms; because of this, they are the most reactive metals.

Metals at the **bottom** of the series are the **least electropositive**, i.e. they **lose** electrons with **difficulty** and therefore do not readily form ions. Their ions are very **unstable** and are easily discharged to form atoms; because of this, they are the least reactive metals.

Metal	Symbol	Reactivity
Potassium	K	Most reactive
Sodium	Na	
Calcium	Ca	
Magnesium	Mg	
Aluminium	Al	Decreasing:
Zinc	Zn	• Ease of ionisation
Iron	Fe	• Reactivity
Lead	Pb	• Stability of compounds
(Hydrogen)	(H)	• Strength as a reducing agent
Copper	Cu	
Mercury	Hg	
Silver	Ag	
Gold	Au	Least reactive

Table 17.1 The reactivity series of metals

REDUCTION REACTIONS BETWEEN METALS AND METAL OXIDES

A metal will **reduce** the oxide of any metal **below** it in the reactivity series:

$$A \ + \ BO \longrightarrow AO + B$$

high in series   lower in series

- A readily ionises forming A^{2+} ions.
- B^{2+} ions are readily discharged forming B atoms.

  e.g. $Mg(s) + CuO(s) \longrightarrow MgO(s) + Cu(s)$

This is also a **displacement** reaction.

DISPLACEMENT REACTIONS BETWEEN METALS AND SOLUTIONS OF METAL SALTS

A metal will **displace** a metal which is **lower** than itself in the reactivity series from a solution of the metal salt:

$$A(s) \ + \ BX(aq) \longrightarrow AX(aq) + B(s)$$

high in series   lower in series

e.g. $Zn(s) + CuSO_4(aq) \longrightarrow ZnSO_4(aq) + Cu(s)$

This is also a **redox** reaction.

Metal	Reaction with air or oxygen	Reaction with water or steam	Reaction with dilute hydrochloric or sulphuric acid	Reduction of metal oxide by: Carbon or carbon monoxide	Hydrogen
K	Burn very easily forming **oxide**, e.g. $2Mg(s) + O_2(g) \longrightarrow 2MgO(s)$	React with **cold water** forming **hydroxide** and **hydrogen**, e.g. $2Na(s) + 2H_2O(l) \longrightarrow 2NaOH(aq) + H_2(g)$	Violent reaction forming **salt** and **hydrogen**	**Not** reduced	**Not** reduced
Na			Violent reaction forming **salt** and **hydrogen**		
Ca			Vigorous reaction forming **salt** and **hydrogen**		
Mg		React with **steam** forming **oxide** and **hydrogen**, e.g. $Zn(s) + H_2O(g) \longrightarrow ZnO(s) + H_2(g)$			
Al	**Burn**, especially if powdered, forming **oxide**		React forming **salt** and **hydrogen**, e.g. $Fe(s) + 2HCl(aq) \longrightarrow FeCl_2(aq) + H_2(g)$		
Zn				Reduced at high temperatures forming **metal** and **carbon dioxide**	
Fe	Do not burn, form **oxide** when **heated strongly**				
Pb		**No reaction**	**No reaction**	Reduced forming **metal** and **carbon dioxide**, e.g. $2PbO(s) + C(s) \longrightarrow 2Pb(s) + CO_2(g)$	Reduced forming **metal** and **steam**, e.g. $CuO(s) + H_2(g) \longrightarrow Cu(s) + H_2O(g)$
Cu					
Ag	**No reaction**				

Table 17.2 The reactions of some metals

Metal	Metal compound		
	Nitrate	**Carbonate**	**Hydroxide**
K	Decompose forming **metal nitrite** and **oxygen**, e.g. $2NaNO_3(s) \xrightarrow{\text{heat}} 2NaNO_2(s) + O_2(g)$	Stable, **not** decomposed by heat	Stable, **not** decomposed by heat
Na			
Ca	Decompose forming **metal oxide**, **nitrogen dioxide** and **oxygen**, e.g. $2Pb(NO_3)_2(s) \xrightarrow{\text{heat}} 2PbO(s) + 4NO_2(g) + O_2(g)$ Ease increases down series	Decompose forming **metal oxide** and **carbon dioxide**, e.g. $CaCO_3(s) \xrightarrow{\text{heat}} CaO(s) + CO_2(g)$ Ease increases down series	Decompose forming **metal oxide** and **water (as steam)**, e.g. $Zn(OH)_2(s) \xrightarrow{\text{heat}} ZnO(s) + H_2O(g)$ Ease increases down series
Mg			
Al			
Zn			
Fe			
Pb			
Cu			
Ag	Decomposes forming **metal**, **nitrogen dioxide** and **oxygen**: $2AgNO_3(s) \xrightarrow{\text{heat}} 2Ag(s) + 2NO_2(g) + O_2(g)$	Unstable	No hydroxide exists

Table 17.3 The stability of metal compounds when heated

The electrochemical series

If two strips of **different** metals are placed in an **electrolyte**, e.g. dilute sulphuric acid, and connected externally through a voltmeter, the voltmeter shows that an **electric current** flows between the metals, i.e. a **chemical cell** is formed.

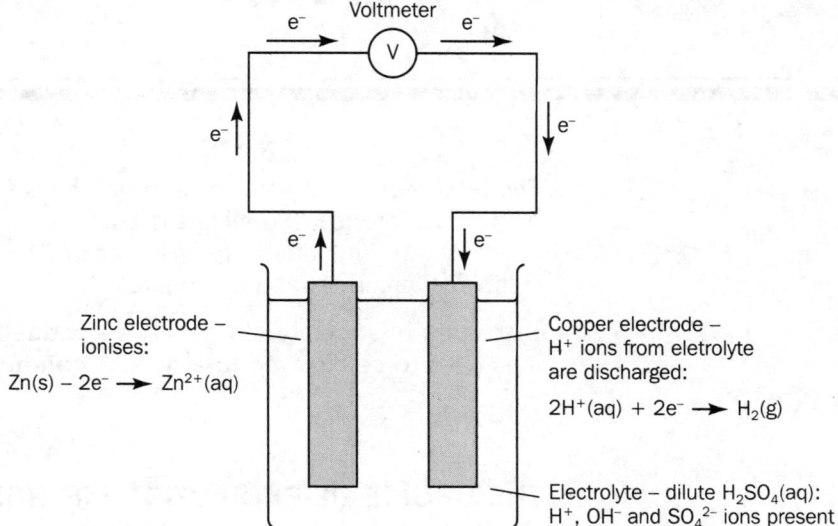

Zinc electrode – ionises:

$Zn(s) - 2e^- \longrightarrow Zn^{2+}(aq)$

Copper electrode – H^+ ions from eletrolyte are discharged:

$2H^+(aq) + 2e^- \longrightarrow H_2(g)$

Electrolyte – dilute $H_2SO_4(aq)$: H^+, OH^- and SO_4^{2-} ions present

Fig. 17.1 A simple chemical cell

- The **more electropositive** of the two metals **ionises** – in Fig. 17.1 **zinc** is more electropositive than copper, therefore zinc ionises forming **Zn^{2+} ions** which dissolve in the electrolyte:

$$Zn(s) - 2e^- \longrightarrow Zn^{2+}(aq)$$

- The **electrons** flow around the circuit to the **copper** strip where **H^+ ions** from the electrolyte are discharged:

$$2H^+(aq) + 2e^- \longrightarrow H_2(g)$$

The **ease** with which each different metal loses electrons can be **compared** with a standard electrode, e.g. a hydrogen electrode. When the metals are then arranged in **order** of ability to lose electrons, they form the **electrochemical series**. The series is identical to the reactivity series, except that **calcium** comes **above sodium**:

K   Ca   Na   Mg   Al   Zn   Fe   Pb   (H)   Cu   Hg   Ag   Au

18 The extraction and uses of metals

The least reactive metals occur in nature as **free elements**, e.g. copper, silver, gold. The more reactive metals occur as **compounds**. These compounds, together with impurities, are called **ores**. The most important ores are oxides, sulphides, chlorides and carbonates.

Extraction of a metal from its ore is a **reduction process** because the metal ions have to be changed to atoms by **gaining** electrons:

$$M^{n+} + ne^- \longrightarrow M$$

THE CHOICE OF EXTRACTION METHOD

The choice of extraction method depends on the **position** of the metal in the reactivity series.

1 **Very reactive metals**: K, Na, Ca, Mg, Al form **stable** ions which are difficult to reduce. They require a **powerful** method of reduction – **electrolysis** of their molten ores. The cathode provides electrons and therefore acts as the reducing agent.
2 **Less reactive metals**: Zn, Fe, Pb form **less stable** ions which are easier to reduce. These need a **less powerful** method of reduction – **heating** the oxide ore with the reducing agents **carbon** (in the form of coke) or **carbon monoxide**.
3 **Least reactive metals**: Cu, Hg, Ag, Au usually occur as **free elements** since their ions are very **unstable**. They may, however, be extracted from their ores by **heating** the ore in **air** at high temperatures; this is the **least powerful** method of reduction.

Extraction of aluminium

Ores: bauxite – impure $Al_2O_3 . 2H_2O$
cryolite – Na_3AlF_6

Extraction is carried out by **electrolysis** in three stages:

1 The bauxite is **purified**.
2 The purified bauxite is **dissolved** in molten cryolite at 900 °C to separate the ions. Electrolysis of molten aluminium oxide is not possible because its melting point is extremely high (2050 °C) and the liquid is a poor conductor.
3 The aluminium oxide/cryolite solution is **electrolysed** using graphite electrodes:

- **At cathode**: $Al^{3+}(l) + 3e^- \longrightarrow Al(l)$

  Molten aluminium collects at the bottom of the cell and is tapped off.

- **At anode**: $2O^{2-}(l) - 4e \longrightarrow O_2(g)$

Oxygen gas collects at the anodes.

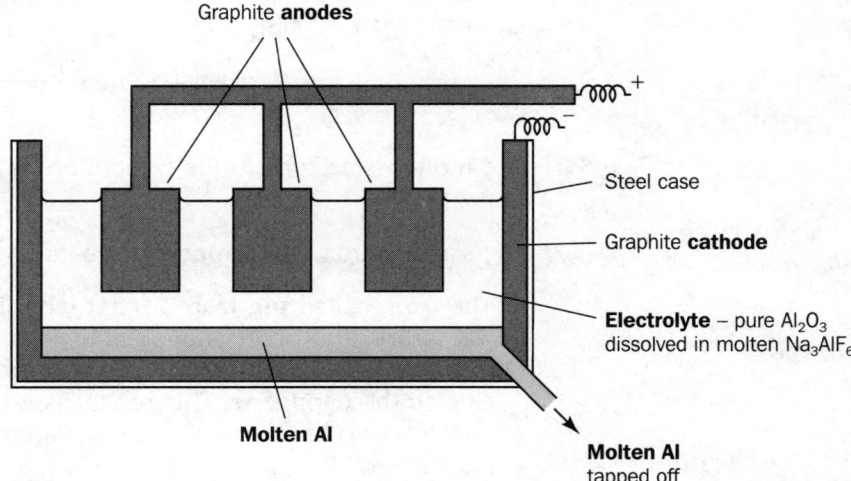

Graphite **anodes**

Steel case

Graphite **cathode**

Electrolyte – pure Al_2O_3 dissolved in molten Na_3AlF_6

Molten Al

Molten Al tapped off

Fig. 18.1 Electrolysis cell for the extraction of aluminium

Extraction of iron

Ores: haematite – Fe_2O_3

magnetite – Fe_3O_4
siderite – $FeCO_3$ } converted to haematite

Extraction is carried out by the **reduction** of haematite by **carbon monoxide** in a **blast furnace**. (See Fig. 18.2 on next page.)

1 A mixture of haematite, limestone ($CaCO_3$) and coke, called the **charge**, is placed into the top of the furnace.

2 **Hot air** is blown into the bottom of the furnace.

3 At **stage 1** (see Fig. 18.2) – coke burns in air forming **carbon dioxide**:

$$C(s) + O_2(g) \longrightarrow CO_2(g) \quad \Delta H - ve$$

Heat produced keeps the temperature at about **1900 °C**.

4 At **stage 2** – more coke reduces the carbon dioxide to **carbon monoxide**:

$$CO_2(g) + C(s) \longrightarrow 2CO(g) \quad \Delta H + ve$$

Heat absorbed reduces the temperature to about **1100 °C**.

5 At **stage 3** – the carbon monoxide reduces the haematite to **iron**:

$$Fe_2O_3(s) + 3CO(g) \longrightarrow 2Fe(s) + 3CO_2(g)$$

The temperature at this part of the furnace is about **700 °C**.
The iron moves down the furnace where it melts (MP 1535 °C).
The **molten iron** then runs to the bottom of the furnace.

6 **Impurities** in the ore, mainly silica (SiO_2), also have to be removed. At temperatures above **850 °C** (between stages 2 and 3) the limestone

decomposes forming **calcium oxide** and more **carbon dioxide**:

$$CaCO_3(s) \longrightarrow CaO(s) + CO_2(g)$$

The silica, being acidic, reacts with the basic calcium oxide to form molten calcium silicate or **slag**:

$$CaO(s) + SiO_2(s) \longrightarrow CaSiO_3(l)$$
$$\text{slag}$$

The **molten slag** runs to the bottom of the furnace where it floats on the iron.

7 The slag and iron are **tapped off** separately:

- The **iron**, called **pig iron** or **cast iron**, is impure, containing about 4% carbon and other impurities, e.g. sulphur and phosphorus. It is further purified by blowing air, enriched with oxygen, through it to oxidise the impurities. This produces **wrought iron**, most of which is converted to **steel** by adding calculated quantities of carbon and other substances.

- The **slag** is used for **road building** or as a **fertiliser**.

8 **Waste gases** (N_2 from the air, H_2, CO, CO_2) are removed from the top of the furnace and **burnt** to heat air blown in at the bottom – conservation.

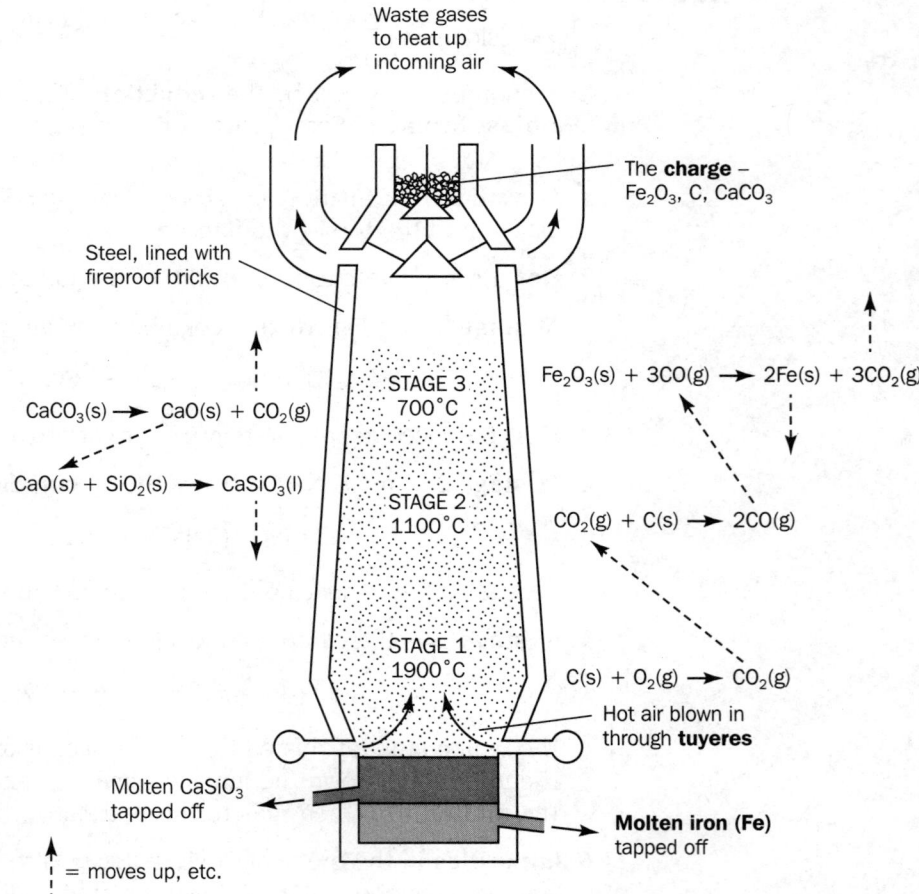

Fig. 18.2 The blast furnace

Corrosion of metals

Corrosion occurs when the surface of a solid, usually a metal, is **'eaten away'** by the action of **oxygen**, **moisture** and **pollutants** present in the atmosphere. When metals corrode, they are initially converted into their **oxides**. Generally, the **higher** a metal is in the reactivity series, the more **rapidly** it corrodes.

CORROSION OF ALUMINIUM

On exposure to the atmosphere, a fresh piece of aluminium forms a layer of **aluminium oxide, Al_2O_3**, which **adheres** to the metal below and protects it against further corrosion.

CORROSION OF IRON AND STEEL – RUSTING

On exposure to the atmosphere, iron and steel corrode to form **rust** which is **hydrated iron(III) oxide, $Fe_2O_3.xH_2O$**. This rust does not adhere to the metal below, instead it **flakes off** exposing fresh metal for further attack. Rust, therefore, gradually 'eats' into the metal.

Methods of preventing rusting

1 Coating the object with **paint**, **grease**, **plastic** or **rubber** which prevent contact with air or water.
2 Coating the object with **zinc**, a process called **galvanising**. Zinc forms an adherent oxide layer which protects against further corrosion and, if damaged, the zinc is oxidised in preference to the iron since it is higher in the electrochemical series.
3 Coating the object with **tin**, e.g. 'tin' cans are made of steel plated with tin. Once scratched, however, rusting occurs rapidly since iron is more reactive than tin.
4 Attaching a **sacrificial anode** to the object, i.e. a piece of a metal higher in the reactivity series, such as magnesium. The magnesium, being higher in the series, is oxidised in preference to the iron.

Uses of metals

The properties of metals determine their uses (see Table 18.1).

Metal	Use	Properties
Aluminium	Cooking utensils	Low density, good conductor of heat, non-toxic, high resistance to corrosion owing to adherent oxide coating, good appearance, inexpensive
	Overhead electrical cables	Low density – cables can be thick but light, very good electrical conductivity
	In metallic paints	Silvery metallic colour gives bright spectacular appearance and reflects heat
Cast iron (impure iron)	Small castings e.g. railings, grates, hot-water pipes	Hard, inexpensive, easy to cast into exact shapes. Brittle, therefore cannot be used where strength is important
Wrought iron (pure iron)	Larger castings, e.g. ornamental metal work, iron sheeting, agricultural implements	Easily hammered and welded when hot, stronger than cast iron, more malleable and less brittle than cast iron

Table 18.1 The uses of selected metals related to their properties

(continued)

Metal	Use	Properties
Lead	Roofing	Very malleable, resistant to corrosion
	In lead-acid accumulators, e.g. car batteries	Good electrical conductor, design of battery (lead plates dipped in dilute $H_2SO_4(aq)$) makes recharging possible
	Protective shield from X-rays and radioactivity	Radiation cannot penetrate through lead

Table 18.1 *Continued*

Alloys

An **alloy** is a **mixture** of two or more metals, a few alloys also contain non-metals. The physical properties of alloys are usually very different from those of the pure metals – in general they are **harder**, **stronger** and more **resistant to corrosion**, therefore are often used in place of pure metals (see Table 18.2).

Alloy	Composition	Use	Properties
Duralumin	95% Al, 4% Cu, traces of Mg and Mn	Aircraft construction	Low density – as light as aluminium, much stronger and more resistant to corrosion than aluminium
Magnalium	70% Al, 30% Mg	As for Duralumin	As for Duralumin
Steel	Fe+ 0.15–1.5% carbon. Other metals may be present, e.g. Cr, Ni, Co, Mn	Car bodies, armour plate, girders, heavy duty machinery, tools, stainless steel cutlery	Harder and much stronger than iron, more resistant to rusting than iron. **Stainless steel** (70% Fe, 20% Cr, 10% Ni) is even harder and more resistant to corrosion than other steels
Solder	67% Pb, 33% Sn	Joining metals, e.g. wires, pipes	Low melting point – lower than lead, harder than lead

Table 18.2 The uses of selected alloys related to their properties

Metals and life

THE IMPORTANCE OF METALS AND THEIR COMPOUNDS TO LIVING ORGANISMS

Many metals, either in the form of **ions** or combined in **compounds**, are essential to plant and animal life.

1 **Iron** forms part of the **haemoglobin** molecule in red blood cells, which is essential to carry **oxygen** in the form of oxyhaemoglobin around the body. Iron is also necessary for the manufacture of the **chlorophyll** molecule in green plants, which is essential for the absorption of sunlight energy for **photosynthesis**.

2 **Magnesium** forms part of the **chlorophyll** molecule.

3 **Calcium** compounds, mainly calcium phosphate, are constituents of **bones** and **teeth**.

4 **Sodium** and **potassium** ions are essential for the transmission of **nerve impulses**.

5 Many other elements, known as **trace elements**, are required in minute quantities by plants and animals, e.g. copper and zinc ions are necessary for the functioning of certain enzymes in cells. Other important trace elements include manganese, molybdenum, chromium and cobalt.

POLLUTION BY METALS AND THEIR COMPOUNDS

Many **heavy metal ions**, especially when combined with organic compounds, are **toxic** to living organisms. They remain in the environment and are **cumulative** poisons which **concentrate** up food chains, affecting higher organisms.

1 **Lead** in compounds from car exhausts affects enzymes, impairing cell metabolism. It can cause brain damage, even death, especially in young children.

2 **Mercury** in compounds from industry causes nervous disorders (e.g. Minamata disease), even death.

3 **Tributyl tin** in antifouling paints harms shellfish.

4 Many others, whilst required in minute quantities by living organisms, become toxic in larger amounts, e.g. zinc, copper and cadmium.

19 The chemistry and uses of some non-metals

A **non-metal** is an element which ionises by **gaining** electrons, i.e. the element is **electronegative**.

General properties of non-metals

(See Table 2.2, p. 6 for the physical properties.)

1 **Non-metals are oxidising agents**.
   When a non-metal reacts and forms ions, it forms **negative ions** by **gaining** electrons from the substance it is reacting with:

$$N + ne^- \longrightarrow N^{n-}$$

2 **Non-metals react with air or oxygen**.
   They form **acidic oxides**. Acidic oxides react with water to form **acids**. A few also form **neutral oxides**, e.g. carbon monoxide, CO, dinitrogen oxide, N_2O.

3 **Non-metals do not react with dilute acids**.

4 **Non-metals do not react with water or steam**.

5 **Non-metals generally react with other non-metals**.
   They form **covalent** compounds.

6 **Non-metals generally react with metals**.
   They form **ionic** compounds.

Hydrogen

PHYSICAL PROPERTIES

1 A colourless, odourless, tasteless gas.
2 Virtually insoluble in water.
3 Less dense than air – it is the lightest gas known.

CHEMICAL PROPERTIES

1 Hydrogen is a **neutral** gas.

2 Hydrogen **burns** in air or oxygen with a blue flame forming **steam**:

$$2H_2(g) + O_2(g) \longrightarrow 2H_2O(g)$$

A mixture of hydrogen and air or oxygen **explodes** when lit. This is the test for hydrogen (see p. 114).

3 Hydrogen reacts with **reactive metals** to form **hydrides**:

e.g. $$Ca(s) + H_2(g) \longrightarrow CaH_2(s)$$

4 Hydrogen is a **reducing agent** – it reduces the oxides of iron and metals below iron in the reactivity series,

e.g. $$PbO(s) + H_2(g) \longrightarrow Pb(s) + H_2O(g)$$

Carbon PHYSICAL PROPERTIES

Carbon has three allotropes: **diamond** and **graphite** (see pp. 28–29), and **amorphous carbon** which has an irregular atomic structure, e.g. charcoal, coke. Charcoal is a brittle, black solid which does not conduct electricity.

CHEMICAL PROPERTIES

1 Carbon **burns** in air or oxygen.

If the oxygen supply is plentiful, **carbon dioxide** is produced:

$$C(s) + O_2(g) \longrightarrow CO_2(g)$$

If the oxygen supply is limited, **carbon monoxide** is produced:

$$2C(s) + O_2(g) \longrightarrow 2CO(g)$$

2 Carbon is a **reducing agent**.

It reduces steam:

$$C(s) + H_2O(g) \longrightarrow CO(g) + H_2(g)$$

It reduces the oxides of zinc and metals below zinc in the reactivity series,

e.g. $$2CuO(s) + C(s) \longrightarrow 2Cu(s) + CO_2(g)$$

	Uses
Diamond, **C**	In **drill tips** In the tips of **cutting tools**, e.g. glass cutters In **jewellery**
Graphite, **C**	As a **lubricant** Mixed with clay forms the 'lead' in **'lead' pencils** To make **electrodes** To **slow down** neutrons in nuclear reactors
Charcoal, **C**	As a **smokeless fuel** In **gas masks** and **water purifiers** to adsorb chemical impurities
Carbon dioxide, **CO₂**	In the manufacture of **fizzy drinks** As a **refrigerant** (dry ice) In **fire extinguishers** (see p. 39)

Table 19.1 Uses of carbon and some of its compounds

(continued)

	Uses
Sodium carbonate, Na_2CO_3	To **soften** hard water In the manufacture of **glass**
Sodium hydrogen-carbonate, $NaHCO_3$	As a **raising agent** in baking powder (see p. 39) As an **antacid** for the treatment of indigestion (see p. 39)

Table 19.1 *Continued*

Nitrogen

PHYSICAL PROPERTIES

1 A colourless, odourless, tasteless gas.

2 Virtually insoluble in water.

3 Slightly less dense than air.

CHEMICAL PROPERTIES

1 Nitrogen is a **neutral** gas.

2 Nitrogen is chemically **inert** under ordinary conditions.

3 Nitrogen reacts with **oxygen** when subjected to a **spark**, forming **nitrogen monoxide**:

$$N_2(g) + O_2(g) \longrightarrow 2NO(g)$$

4 Nitrogen reacts with **hydrogen** under pressure and in the presence of a catalyst forming **ammonia**:

$$N_2(g) + 3H_2(g) \rightleftharpoons 2NH_3(g)$$

5 Nitrogen reacts with **reactive metals** at high temperatures forming **nitrides**:

e.g. $$3Mg(s) + N_2(g) \longrightarrow Mg_3N_2(s)$$

	Uses
Nitrogen, N_2	As a **refrigerant** In **food packaging** to provide an inert atmosphere which prevents aerobic decay In the manufacture of **ammonia** (see p. 117)
Ammonia, NH_3	In the manufacture of **fertilisers**, e.g. NH_4NO_3, $(NH_4)_2SO_4$, $(NH_4)_2HPO_4$ In the manufacture of **nitric acid** (see p. 118) and **plastics**
Nitric acid, HNO_3	In the manufacture of **fertilisers**, e.g. NH_4NO_3 In the manufacture of **explosives** and **dyes**

Table 19.2 Uses of nitrogen and some of its compounds

Oxygen

PHYSICAL PROPERTIES

1 A colourless, odourless, tasteless gas.

2 Slightly soluble in water.

3 Slightly denser than air.

CHEMICAL PROPERTIES

1 Oxygen is a **neutral** gas.

2 Oxygen is a powerful **oxidising agent** – any reaction involving oxygen is **oxidation**.

3 Oxygen **supports combustion** – many substances burn in air or oxygen, the reactions being **exothermic**.

4 Most metals and non-metals combine with oxygen to form **oxides** (see p. 91 and below).

5 Oxygen causes **rusting** – rust is hydrated iron(III) oxide, $Fe_2O_3 . xH_2O$, which only forms when both oxygen (air) and water attack iron.

6 Oxygen is essential for **aerobic respiration** in all living cells:

$$C_6H_{12}O_6(aq) + 6O_2(g) \longrightarrow 6CO_2(g) + 6H_2O(l) + energy$$

OXIDES

Oxides can be classified into **five** groups:

1 Acidic oxides

These are oxides of **non-metals** which react with water to form **acids**,

e.g.
$$SO_2(g) + H_2O(l) \rightleftharpoons H_2SO_3(aq)$$
$$SO_3(g) + H_2O(l) \longrightarrow H_2SO_4(aq)$$

Acidic oxides react with alkalis forming a **salt** and **water** only,

e.g.
$$CO_2(g) + 2NaOH(aq) \longrightarrow Na_2CO_3(aq) + H_2O(l)$$

2 Basic oxides

These are oxides of **metals** which react with acids forming a **salt** and **water** only,

e.g.
$$CuO(s) + H_2SO_4(aq) \longrightarrow CuSO_4(aq) + H_2O(l)$$

Potassium oxide, K_2O, sodium oxide, Na_2O, and calcium oxide, CaO, react with water forming **alkalis**,

e.g.
$$K_2O(s) + H_2O(l) \longrightarrow 2KOH(aq)$$

3 Amphoteric oxides

These are oxides of **aluminium**, **zinc** and **lead** which show basic and acidic properties, i.e. they react with both **acids** and **strong alkalis**,

e.g. **PbO** as a **basic** oxide:

$$PbO(s) + 2HNO_3(aq) \longrightarrow Pb(NO_3)_2(aq) + H_2O(l)$$

PbO as an **acidic** oxide:

$$2NaOH(aq) + PbO(s) \longrightarrow Na_2PbO_2 + H_2O(l)$$
sodium
plumbate(II)

4 Neutral oxides

These are oxides of **certain non-metals** which do not react with either an acid or a base, e.g. carbon monoxide, CO, nitrogen monoxide, NO, dinitrogen oxide, N_2O.

5 Peroxides

These are oxides of reactive metals which contain the O_2^{2-} ion, i.e. they contain **more** oxygen than the basic oxide. Peroxides produce **hydrogen peroxide** when treated with dilute acids,

e.g. $$Na_2O_2(s) + H_2SO_4(aq) \longrightarrow Na_2SO_4(aq) + H_2O_2(aq)$$

Sulphur

PHYSICAL PROPERTIES

Sulphur has two main allotropes, **rhombic sulphur** and **monoclinic sulphur**. Both are yellow, crystalline solids composed of S_8 molecules. The crystals of rhombic sulphur are **octahedral** in shape whilst those of monoclinic sulphur are **needle-shaped**.

CHEMICAL PROPERTIES

1 Sulphur **burns** in air or oxygen with a blue flame forming **sulphur dioxide**:

$$S(s) + O_2(g) \longrightarrow SO_2(g)$$

2 Sulphur reacts with **metals** to form **sulphides**:

e.g. $$Mg(s) + S(s) \longrightarrow MgS(s)$$

	Uses
Sulphur, **S**	To **vulcanise** (harden) rubber, e.g. for car tyres In the manufacture of **medicinal drugs** and **ointments** for the treatment of fungal infections In the manufacture of **fungicides** for agricultural use In the manufacture of **gunpowder** and **matches** In the manufacture of **sulphur dioxide** and **sulphuric acid** (see p. 116)
Sulphuric acid, **H_2SO_4**	In the manufacture of **fertilisers**, e.g. $(NH_4)_2SO_4$ In the manufacture of **paint pigments**, **dyes**, **detergents** and **plastics**
Sulphur dioxide, **SO_2**	As a **food preservative**, e.g. in jams, fruit juices As a **bleaching agent** in the manufacture of paper

Table 19.3 Uses of sulphur and some of its compounds

Chlorine

PHYSICAL PROPERTIES

1 A yellow-green, poisonous gas with a pungent, choking smell.

2 Moderately soluble in water.

3 Denser than air.

CHEMICAL PROPERTIES

1 Chlorine is an **acidic** gas.

2 Chlorine reacts with **metals** forming **anhydrous ionic chlorides**:

e.g. $$Zn(s) + Cl_2(g) \longrightarrow ZnCl_2(s)$$

3 Chlorine reacts with some **non-metals** forming **covalent chlorides**:

e.g. $$H_2(g) + Cl_2(g) \longrightarrow 2HCl(g)$$

4 Chlorine is a powerful **oxidising agent** – chlorine atoms readily **gain** electrons to form **stable Cl^- ions**:

$$Cl_2 + 2e^- \longrightarrow 2Cl^-$$

5 In the presence of moisture, chlorine acts as a **bleaching agent.** Chlorine combines with water forming **hydrochloric acid** and **chloric(I) acid**:

$$Cl_2(g) + H_2O(l) \longrightarrow HCl(aq) + \underset{\text{chloric(I) acid}}{HClO(aq)}$$

Chloric(I) acid is the bleaching agent; it bleaches by **oxidising** certain coloured substances to their colourless form:

$$HClO(aq) + \underset{\text{(coloured)}}{\text{dye}} \longrightarrow HCl(aq) + \underset{\text{(colourless)}}{\text{oxidised dye}}$$

6 Chlorine reacts with **alkalis** forming **chlorides** and **chlorate(I) salts**:

e.g. $$2NaOH(aq) + Cl_2(g) \longrightarrow NaCl(aq) + NaClO(aq) + H_2O(l)$$

7 Chlorine displaces **halogens below it** in Group VII from their compounds:

e.g. $$2KBr(aq) + Cl_2(g) \longrightarrow 2KCl(aq) + Br_2(aq)$$

	Uses
Chlorine, **Cl$_2$**	In the manufacture of **sodium chlorate(I)**, **calcium chlorate(I)** and many **chlorinated organic compounds**
Sodium chlorate(I), **NaClO** and calcium chlorate(I) **Ca(ClO)$_2$**	As **bleaching** agents To **sterilise** drinking and swimming pool water – compounds release 'free chlorine' which is toxic to bacteria
Chlorinated organic compounds	As **insecticides**, e.g. DDT As **antiseptics**, e.g. TCP and Dettol In **de-greasing** agents and **dry-cleaning** fluids As **aerosol propellants**, e.g. chlorofluorocarbons

Table 19.4 Uses of chlorine and some of its compounds

Silicon and phosphorus

	Uses
Silicon, **Si**	In the manufacture of **silicon chips** for computers
Silica (quartz) **SiO$_2$**	In **jewellery**, e.g. amethyst is purple quartz In the manufacture of **glass**
Silicates, **SiO$_3{}^{2-}$**	In the manufacture of **ceramic products**

Table 19.5 Uses of silicon and some of its compounds

	Uses
Phosphorus, **P**	In the manufacture of **phosphoric acid**
Calcium phosphate, **Ca$_3$(PO$_4$)$_2$**	In the manufacture of **phosphate fertilisers**, e.g. superphosphate, Ca(H$_2$PO$_4$)$_2$/CaSO$_4$
Phosphoric acid, **H$_3$PO$_4$**	In the manufacture of **phosphate fertilisers**, e.g. (NH$_4$)$_2$HPO$_4$ To **rust-proof** iron and steel
Phosphorus sulphides, e.g. **P$_2$S$_5$**	In the manufacture of **matches**

Table 19.6 Uses of phosphorus and some of its compounds

20 Non-metals in living systems and the environment

Natural cycles and recycling

Elements and compounds are continually being **recycled** in nature. Many pass through the bodies of **living organisms**, are released and re-used by other living organisms. Others are released, by **decomposers**, from the bodies of organisms when they **die**. These can then be re-used by living organisms. This recycling is essential for life to continue.

Man is also constantly removing **natural resources** from the Earth, many of which are non-renewable. These resources are then either burned to release **energy** or used and eventually **discarded**. To conserve our natural resources, **alternative energy sources** must be used increasingly and waste products must be **recycled**, e.g. paper, glass, plastics, metals such as iron, steel, tin, copper, lead and aluminium, industrial waste such as solvents, human and agricultural waste such as vegetable peelings and bagasse.

The water cycle

Water is continually being recycled in nature. This recycling:

1 Ensures a constant supply of water is available to plants for **photosynthesis**.
2 Ensures a constant supply of water is available to all living organisms to keep their cells **hydrated** and to act as a **solvent**.
3 Ensures aquatic organisms have a constant **environment** in which to live.

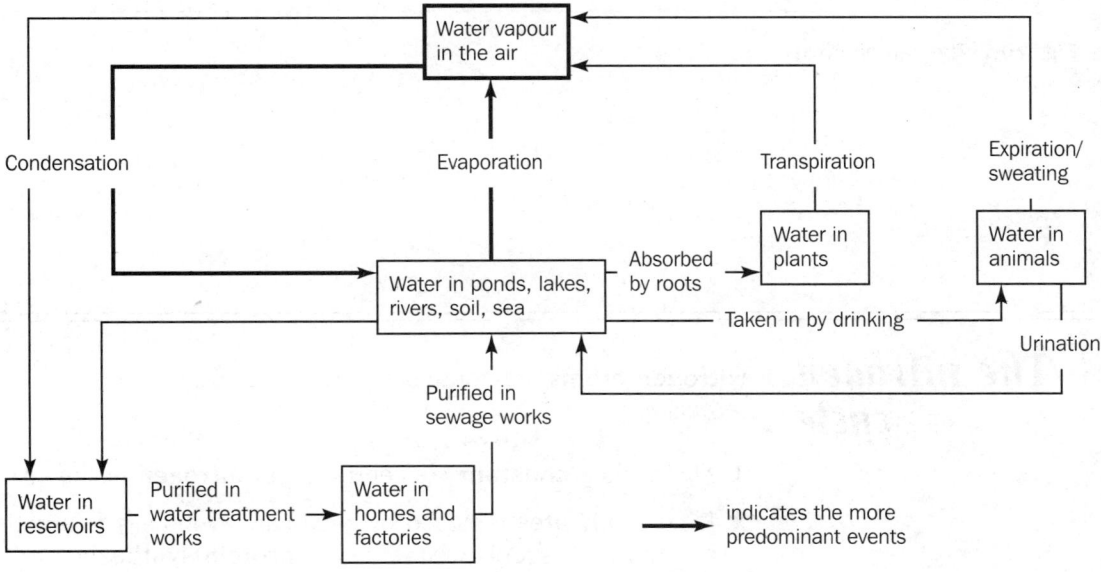

Fig. 20.1 The water cycle

The carbon cycle

Carbon atoms are continually being cycled in nature. This recycling:

1 Maintains the concentration of **carbon dioxide** in the air at an almost constant level, thus ensuring a constant supply of carbon dioxide is available to green plants for **photosynthesis** and marine animals for **skeleton** formation.

2 Ensures a constant supply of **organic food** is available to animals and decomposers.

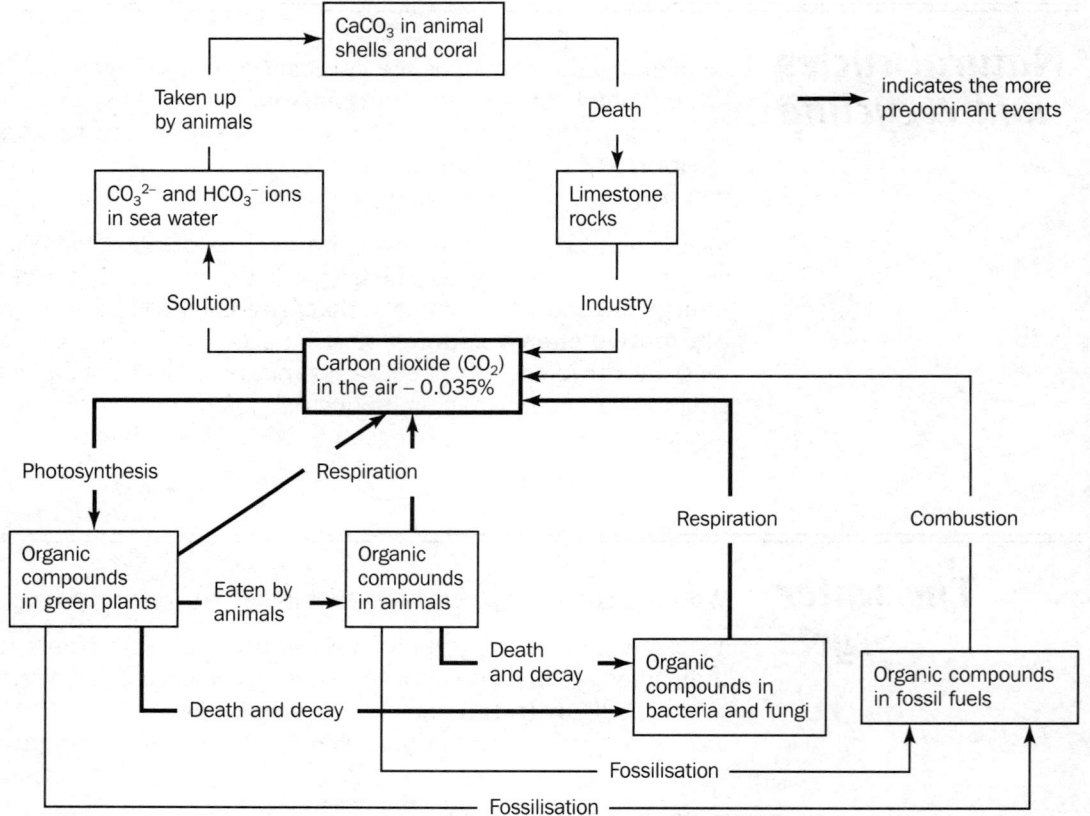

Fig. 20.2 The carbon cycle

The nitrogen cycle

Nitrogen atoms are continually being cycled in nature (see Fig. 20.3). This recycling:

1 Maintains a **constant** concentration of **nitrogen** in the air.

2 Ensures **nitrates** removed by plants are eventually returned, thus plants have a continuous supply of nitrates for **protein synthesis**.

3 Ensures animals and decomposers have a continual supply of **protein**.

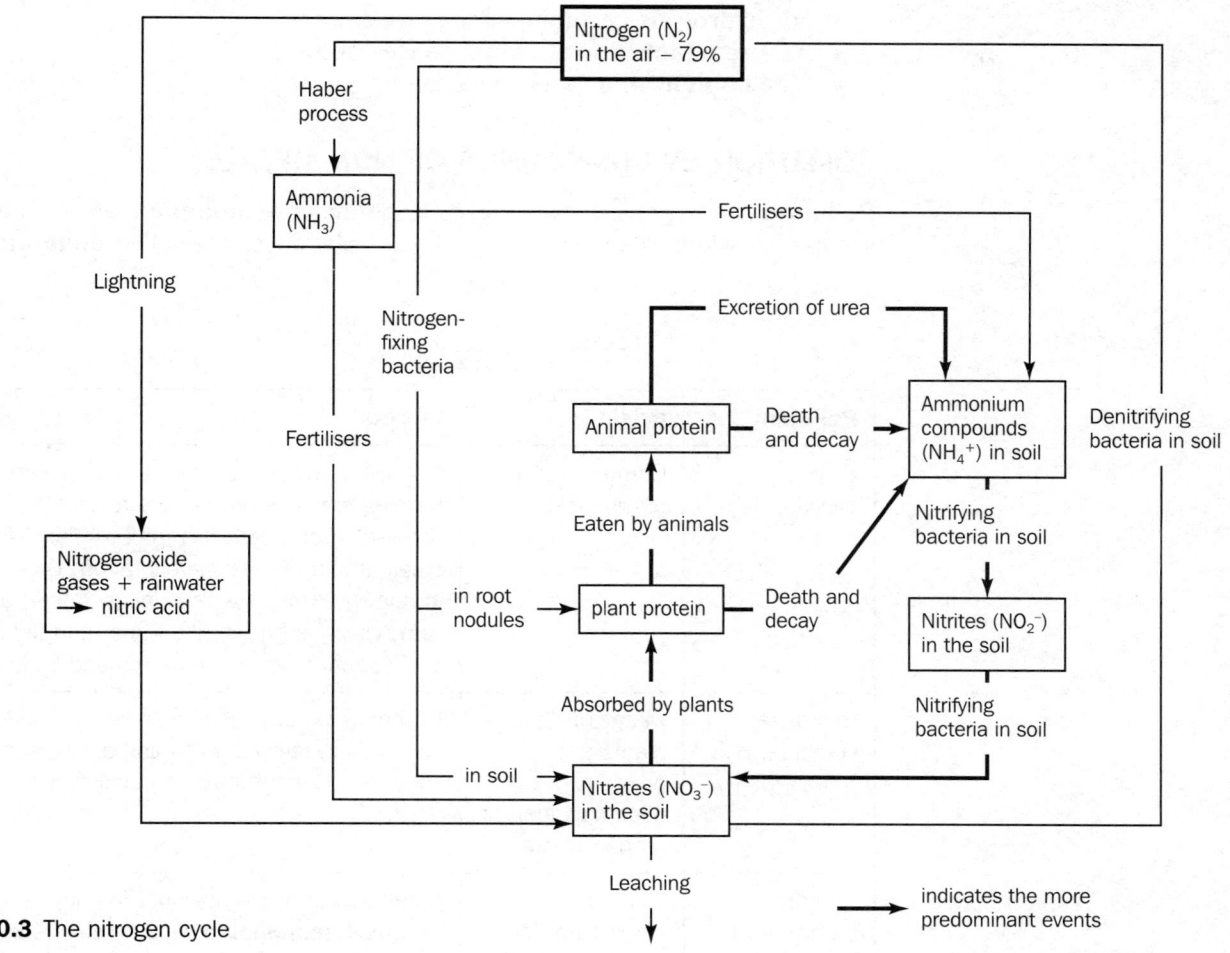

Fig. 20.3 The nitrogen cycle

**Non-metals
and life**

THE IMPORTANCE OF NON-METALS AND THEIR COMPOUNDS TO LIVING ORGANISMS

1 **Carbon dioxide** is essential for **photosynthesis**:

$$6CO_2(g) + 6H_2O(l) \xrightarrow[\text{chlorophyll}]{\text{sunlight}} C_6H_{12}O_6(aq) + 6O_2(g)$$
$$\text{glucose}$$

2 **Oxygen** is essential for **aerobic respiration**:

$$C_6H_{12}O_6(aq) + 6O_2(g) \longrightarrow 6CO_2(g) + 6H_2O(l) + \text{energy}$$

3 **Nitrogen**, obtained as the NO_3^- ion, and **sulphur**, obtained as the SO_4^{2-} ion, are essential for **protein synthesis** in plants.

4 **Phosphorus**, obtained as the PO_4^{3-} ion, is essential for building strong **bones** and **teeth**.

5 **Water** is essential:

- As a **solvent** to dissolve body chemicals and enzymes so they can react, waste products so they can be excreted, digested food so it can be absorbed, and a variety of substances to be transported around the body.

109

- For **hydrolysis** occurring during digestion.
- As a **reactant** in photosynthesis (see above).
- As a **coolant** when it evaporates, e.g. sweat.

POLLUTION BY COMPOUNDS OF NON-METALS

Pollution is any process which leads to an increase in **unpleasant** or **harmful** substances within the environment. These substances are called **pollutants**.

Pollutant	Origin	Effects
Sulphur dioxide, SO_2	Combustion of fossil fuels	Combines with water vapour and smoke forming **smog** which causes respiratory disease, even death. Harmful to plant life – small amounts destroy vegetation. Dissolves in rainwater forming an acidic solution, **acid rain**, which is harmful to plants and animals and corrodes metal structures and buildings
Hydrogen sulphide, H_2S	Decay of dead plants and animals, decomposition of human waste	Extremely toxic, causes respiratory disease, dissolves in rainwater forming a weakly acidic solution, i.e. contributes to acid rain
Carbon monoxide, CO	Incomplete combustion of fossil fuels, e.g. in motor vehicles	Combines with haemoglobin forming **carboxyhaemoglobin** which prevents blood carrying oxygen, resulting in headaches, unconsciousness, death
Carbon dioxide, CO_2	Combustion of fossil fuels	Builds up in atmosphere trapping heat – the **'greenhouse effect'**. This will cause atmospheric temperatures to rise resulting in ice caps melting and extensive flooding
Oxides of nitrogen, NO, NO_2	Combustion at high temperatures in motor vehicles	Extremely toxic, cause respiratory disease, dissolve in rainwater forming an acidic solution i.e. contribute to acid rain
Smoke (carbon particles)	Burning coal	Causes respiratory disease, increases risk of lung cancer, blackens trees reducing photosynthesis, blackens buildings, forms smog
Chloro-fluorocarbons	Aerosol propellants	Break down **ozone layer** of outer atmosphere allowing more ultraviolet light to reach the Earth; this increases mutations and skin cancer

Table 20.1 Air pollutants

Pollutant	Origin	Effects
Organic waste	Untreated sewage, farmyard waste	Aerobic bacteria decompose waste, multiply and use up dissolved oxygen, aquatic organisms die, anaerobic disease-causing bacteria increase, water becomes stagnant and smelly
NO_3^-, PO_4^{3-}, SO_4^{2-} ions	Untreated sewage, synthetic detergents, fertilisers	Cause **eutrophication** (rapid growth of green algae), water turns green, algae die, aerobic bacteria multiply and use up dissolved oxygen (as above). Detergents are also toxic to aquatic organisms and form foam on water which prevents oxygen dissolving
Insecticides, e.g. DDT, and herbicides	Washed off land	Remain in environment and concentrate up food chains harming, even killing, top consumers
Oil	Spilled from tankers and offshore rigs	Forms slicks on sea which prevent oxygen dissolving, toxic to aquatic organisms, smothers birds preventing flight, clogs respiratory systems, ruins beaches

Table 20.2 Water pollutants

Laboratory preparation and identification of gases

Preparation of carbon dioxide and oxygen

Gases can be prepared in the laboratory using various apparatus:

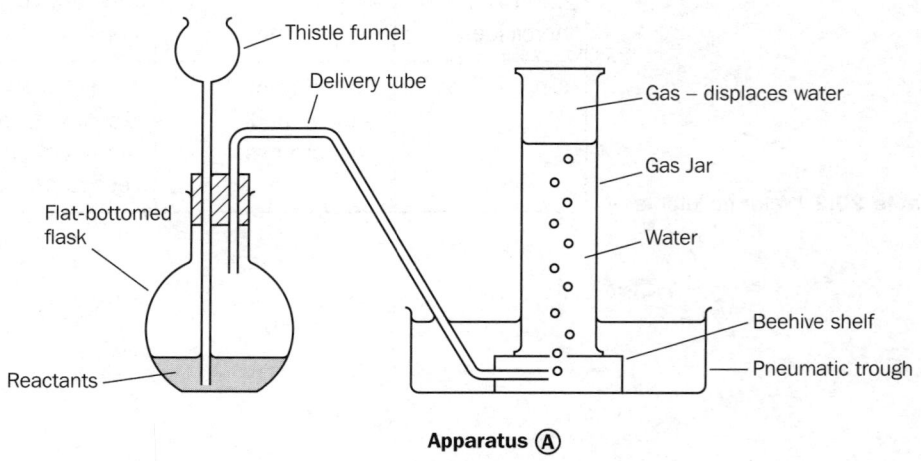

Apparatus Ⓐ

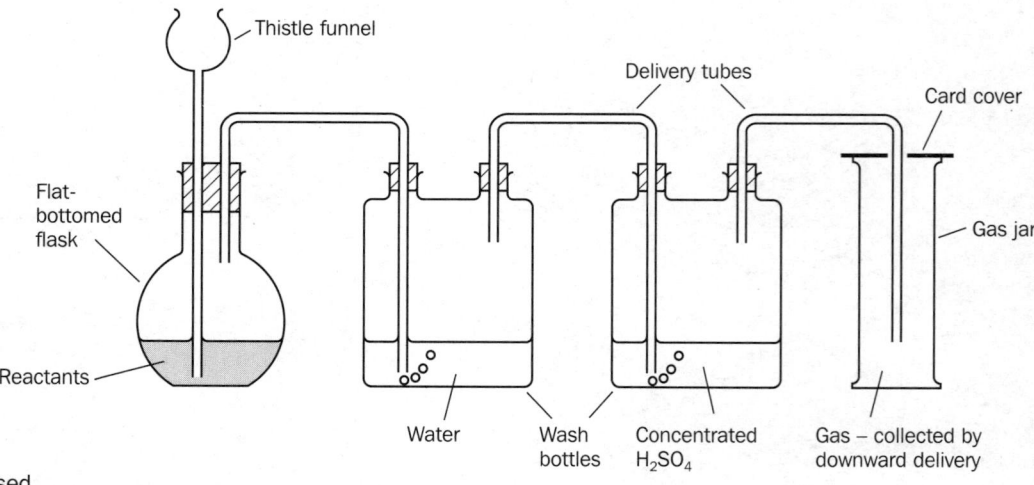

Fig. 21.1 Apparatus used for laboratory preparation of carbon dioxide and oxygen

Apparatus Ⓑ

	Carbon dioxide, CO_2	Oxygen, O_2
Method	Dilute acid + carbonate (not $H_2SO_4 + CaCO_3$). Usually dilute HCl and $CaCO_3$: $$CaCO_3(s) + 2HCl(aq) \longrightarrow$$ $$CaCl_2(aq) + H_2O(l) + CO_2(g)$$	Decomposition of hydrogen peroxide. MnO_2 is used as a catalyst: $$2H_2O_2(aq) \xrightarrow{MnO_2} 2H_2O(l) + O_2(g)$$
Drying agent	Concentrated H_2SO_4 (or anhydrous $CaCl_2$)	Concentrated H_2SO_4 (or anhydrous $CaCl_2$)
Apparatus	For **wet CO_2**: apparatus Ⓐ For **dry CO_2**: apparatus Ⓑ – water removes hydrogen chloride fumes from the acid. Gas collected by downward delivery – denser than air	For **wet O_2**: apparatus Ⓐ For **dry O_2**: apparatus Ⓑ omitting wash bottle containing water. Gas collected by downward delivery – slightly denser than air

Table 21.1 Laboratory preparation of carbon dioxide and oxygen

Preparation of ammonia

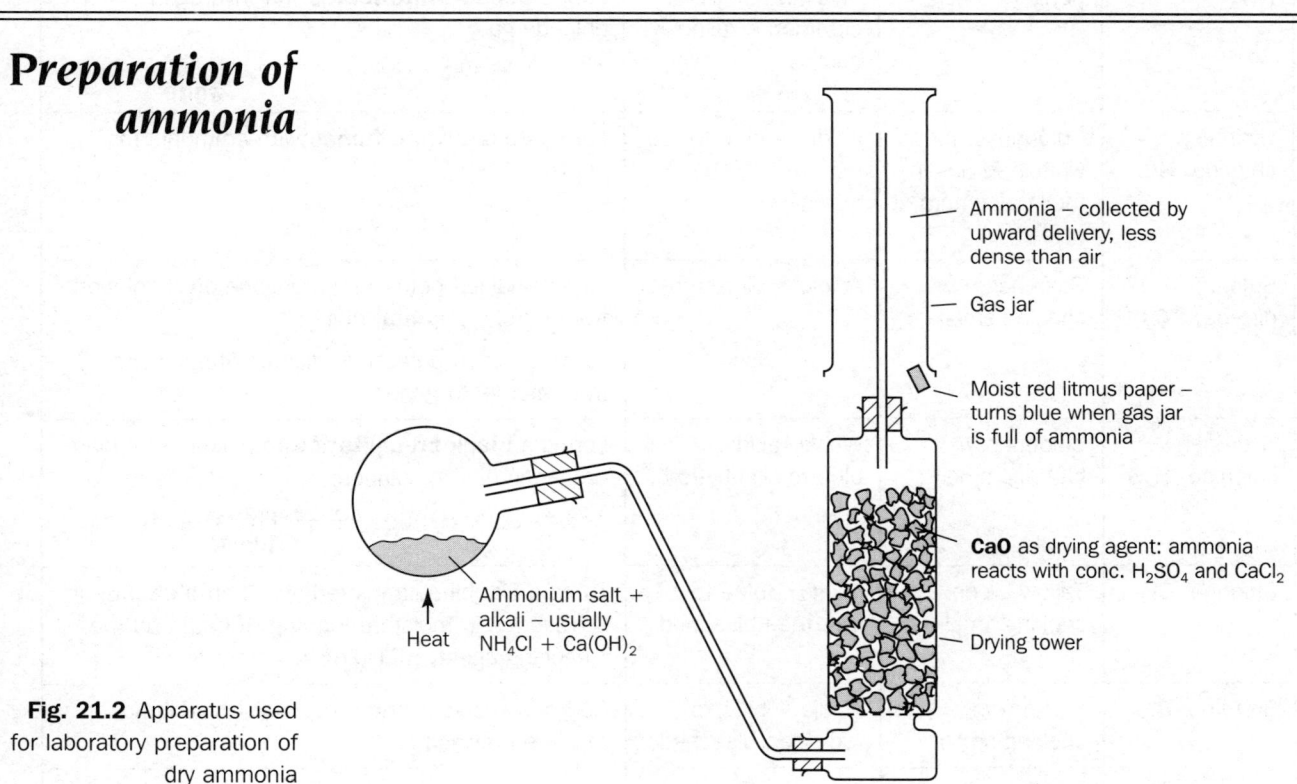

Ammonia – collected by upward delivery, less dense than air

Gas jar

Moist red litmus paper – turns blue when gas jar is full of ammonia

CaO as drying agent: ammonia reacts with conc. H_2SO_4 and $CaCl_2$

Drying tower

Ammonium salt + alkali – usually $NH_4Cl + Ca(OH)_2$

Heat

Fig. 21.2 Apparatus used for laboratory preparation of dry ammonia

N.B. Ammonia is soluble in water, therefore cannot be collected over water.

Identification of gases

Gas	Colour/odour	Effect on moist litmus	Method of identification
Oxygen, O_2	Colourless/ odourless	Neutral – no effect	**Relights** a glowing splint
Hydrogen, H_2	Colourless/ odourless	Neutral – no effect	Lighted splint **'pops'** if air is also present
Carbon dioxide, CO_2	Colourless/ odourless	Weakly acidic – blue to slightly red	A **white precipitate** forms in limewater $Ca(OH)_2(aq) + CO_2(g) \longrightarrow CaCO_3(s) + H_2O(l)$ **white precipitate** Precipitate redissolves on continued bubbling: $CaCO_3(s) + H_2O(l) + CO_2(g) \longrightarrow$ $Ca(HCO_3)_2(aq)$
Ammonia, NH_3	Colourless/ pungent smell	Alkaline – red to blue (the only common alkaline gas)	Turns moist red litmus **blue**. Forms **dense white fumes** with hydrogen chloride gas: $NH_3(g) + HCl(g) \longrightarrow NH_4Cl(s)$ **white**
Hydrogen chloride, **HCl**	Colourless, forms whitish fumes in moist air/sharp, acid smell	Acidic – blue to red	Forms **dense white fumes** with ammonia (as above)
Sulphur dioxide, SO_2	Colourless/ choking smell	Acidic – blue to red	Turns acidified potassium manganate(VII) solution from purple to **colourless** Turns acidified potassium dichromate(VI) solution from orange to **green**
Hydrogen sulphide, H_2S	Colourless/ bad egg smell	Weakly acidic – blue to slightly red	Forms a **black precipitate** with lead(II) ethanoate or lead(II) nitrate solution: $Pb^{2+}(aq) + H_2S(g) \longrightarrow PbS(s) + 2H^+(aq)$ **black**
Chlorine, Cl_2	Yellow-green/ choking smell	Acidic – blue to red, then bleached	Turns moist blue litmus red and then **bleaches** it – dissolves in moisture forming HCl(aq) and the bleaching agent, HClO(aq)
Bromine, Br_2	Red-brown/ choking smell	Acidic – blue to red, then bleached	Red-brown colour **and** turns moist blue litmus red, then **bleaches** it
Nitrogen dioxide, NO_2	Red-brown/ irritating smell	Acidic – blue to red	Red-brown colour **and** turns moist blue litmus **red** but does not bleach it
Water vapour, H_2O	Colourless/ odourless	Neutral – no effect	Turns anhydrous cobalt(II) chloride from blue to **pink** Turns anhydrous copper(II) sulphate from white to **blue**

Table 21.2 Identification of gases

22 Some industrial processes

Manufacture of chlorine and sodium hydroxide

Chlorine and **sodium hydroxide** are manufactured by **electrolysis of brine** (concentrated sodium chloride solution) using a **graphite anode** and a **mercury cathode**.

Ions present: From H_2O: $H^+(aq)$, $OH^-(aq)$
From NaCl: $Na^+(aq)$, $Cl^-(aq)$

1 **Manufacture of chlorine**
At anode: Cl^- ions are discharged as they are present in a higher concentration than OH^-:

$$2Cl^-(aq) - 2e^- \longrightarrow Cl_2(g)$$

Chlorine gas is collected.

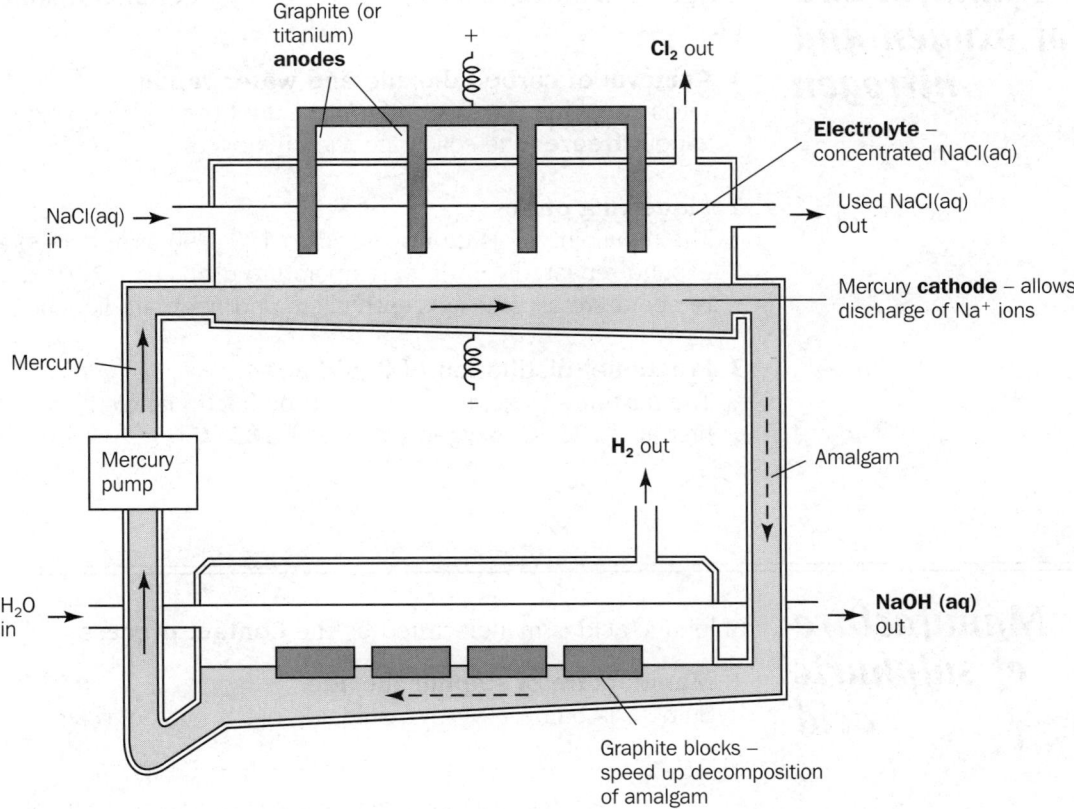

Fig. 22.1 Flowing mercury cathode cell for the manufacture of chlorine and sodium hydroxide

2 Manufacture of sodium hydroxide

At cathode: Na^+ ions are discharged when the cathode is **mercury**:

$$Na^+(aq) + e^- \longrightarrow Na(s)$$

The sodium immediately dissolves in the mercury forming an **amalgam**:

$$Na(s) + Hg(l) \longrightarrow \underset{\text{amalgam}}{Na/Hg(l)}$$

The amalgam is mixed with water, producing **sodium hydroxide solution**, **hydrogen** and **pure mercury** (which is recycled):

$$2Na/Hg(l) + 2H_2O(l) \longrightarrow 2NaOH(aq) + H_2(g) + 2Hg(l)$$

MANUFACTURE OF BLEACH

Sodium chlorate(I) (a bleach) is made by passing chlorine through sodium hydroxide solution:

$$2NaOH(aq) + Cl_2(g) \longrightarrow NaCl(aq) + \underset{\text{sodium chlorate(I)}}{NaClO(aq)} + H_2O(l)$$

Sodium chlorate(I) is a powerful **oxidising agent** which oxidises many coloured dyes to their colourless form.

Manufacture of oxygen and nitrogen

Oxygen and **nitrogen** are manufactured by **fractional distillation of liquid air**.

1 Removal of carbon dioxide and water vapour
Air is cooled in a refrigeration plant until the carbon dioxide and water vapour **freeze**. The solids are then removed.

2 Liquefying of air
The remaining air is **compressed** to 150–200 atm, cooled and allowed to expand repeatedly until its temperature drops to **−200 °C**. At this temperature all gases except helium and neon are liquefied.

3 Fractional distillation of liquid air
The mixture of liquids is separated by fractional distillation – **nitrogen** distils first at **−196 °C**, **oxygen** distils at **−183 °C**.

Manufacture of sulphuric acid

Sulphuric acid is manufactured by the **Contact process**.

1 Manufacture of sulphur dioxide
Sulphur is burnt in air or a sulphide ore is heated in air:

$$S(s) + O_2(g) \longrightarrow SO_2(g)$$

or $\qquad 2ZnS(s) + 3O_2(g) \longrightarrow 2ZnO(s) + 2SO_2(g)$

The sulphur dioxide is then purified.

2 Manufacture of sulphur trioxide

Sulphur dioxide and air are passed over a vanadium(v) oxide (V_2O_5) catalyst at 450 °C and 2 atm pressure:

$$2SO_2(g) + O_2(g) \underset{\text{450 °C, 2 atm}}{\overset{\text{V}_2\text{O}_5 \text{ catalyst}}{\rightleftharpoons}} 2SO_3(g)$$

3 Manufacture of sulphuric acid

Sulphur trioxide is dissolved in concentrated sulphuric acid to form **oleum**:

$$SO_3(g) + H_2SO_4(l) \longrightarrow \underset{\text{oleum}}{H_2S_2O_7(l)}$$

The oleum is diluted with water forming **concentrated sulphuric acid (98% H_2SO_4)**:

$$H_2S_2O_7(l) + H_2O(l) \longrightarrow 2H_2SO_4(l)$$

Sulphur trioxide cannot be added directly to water because the reaction is so extremely **exothermic** that the sulphuric acid vaporises forming a highly corrosive mist.

Manufacture of ammonia

Ammonia is manufactured by the **Haber process**.

1 Manufacture of hydrogen

Hydrogen is made from natural gas by **steam reforming**. Methane and steam are passed over a nickel catalyst at high temperature and pressure:

$$\underset{\text{methane}}{CH_4(g)} + H_2O(g) \underset{\text{100 °C, 50 atm}}{\overset{\text{Ni catalyst}}{\longrightarrow}} CO(g) + 3H_2(g)$$

The gaseous products are mixed with more steam and passed over an iron(III) oxide catalyst:

$$CO(g) + 3H_2(g) + H_2O(g) \overset{\text{Fe}_2\text{O}_3 \text{ catalyst}}{\longrightarrow} CO_2(g) + 4H_2(g)$$

The carbon dioxide is removed by dissolving it in water under pressure.

2 Manufacture of nitrogen

Nitrogen is made by **fractional distillation** of liquid air (see p. 116) or by **burning hydrogen** in air:

$$2H_2(g) + \underset{\text{air}}{\underline{O_2(g) + N_2(g)}} \longrightarrow 2H_2O(l) + N_2(g)$$

3 Manufacture of ammonia

Dry nitrogen and hydrogen are mixed in volume proportions of 1 : 3, pressurised to 200 atm and passed over an iron catalyst at 500 °C

$$N_2(g) + 3H_2(g) \underset{\text{500 °C, 200 atm}}{\overset{\text{Fe catalyst}}{\rightleftharpoons}} 2NH_3(g)$$

The ammonia is **liquefied** by cooling under pressure, and separated from any unreacted gases, which are **recycled**.

Manufacture of nitric acid

Nitric acid is manufactured by the **oxidation of ammonia**.

1 Manufacture of nitrogen monoxide (NO)

Cleaned ammonia and excess air are compressed to 4 atm and passed over a platinum (90%)/rhodium (10%) gauze catalyst heated to red heat:

$$4NH_3(g) + 5O_2(g) \xrightarrow[\text{900 °C, 4 atm}]{\text{Pt/Rh catalyst}} 4NO(g) + 6H_2O(g) \qquad \Delta H - ve$$

The reaction is **exothermic**, and once started it maintains the temperature of the catalyst at about 900 °C.

2 Manufacture of nitric acid

Air, water and nitrogen monoxide are passed into an **absorption tower** at 10 atm. Nitrogen monoxide is oxidised to nitrogen dioxide which reacts with water and more oxygen forming **'concentrated' nitric acid (68% HNO$_3$)**:

$$2NO(g) + O_2(g) \longrightarrow 2NO_2(g)$$

$$4NO_2(g) + 2H_2O(l) + O_2(g) \longrightarrow 4HNO_3(aq)$$

The acid can be further concentrated by **distillation**.

23 Qualitative analysis

Qualitative analysis involves **identifying** the constituents of single substances or mixtures of substances.

Appearance

Preliminary examination of the **appearance** of a substance can indicate what it **might** contain.

Observation	Indication
Black in colour	Oxide (O^{2-}) or sulphide (S^{2-})
Blue or **blue-green** in colour	Copper(II) (Cu^{2+}) salt
Pale green in colour	Iron(II) (Fe^{2+}) salt
Yellow-brown in colour	Iron(II) (Fe^{3+}) salt
Ammonia smell	Ammonium (NH_4^+) salt
Deliquescent (absorbs water, eventually dissolving)	Chloride (Cl^-) or nitrate (NO_3^-)

Table 23.1 The appearance of substances

Tests to identify cations

1 FLAME TEST

Moisten a small amount of the solid with concentrated hydrochloric acid. Dip a clean piece of nichrome or platinum wire into the mixture and heat strongly in a bunsen burner flame.

Flame colour	Ion indicated
Lilac	K^+
Persistent **orange-yellow**	Na^+
Brick red	Ca^{2+}
Bluish	Pb^{2+}
Blue-green	Cu^{2+}

Table 23.2 Flame tests

2 EFFECT OF DILUTE SODIUM HYDROXIDE SOLUTION

a Prepare a **solution** of the solid.
b Add dilute sodium hydroxide solution **dropwise**. Look for a **precipitate**.
c Add **excess** sodium hydroxide solution. Look for the precipitate **redissolving**.
d If **no** precipitate forms in Step b, **warm** gently and test for **ammonia**.

Effect of dilute sodium hydroxide solution		Cation
Dropwise	**Excess**	
White precipitate	Precipitate **insoluble**	Ca^{2+}
White precipitate	Precipitate **soluble** – dissolves forming colourless solution	Al^{3+}, Pb^{2+}, Zn^{2+}
Blue precipitate	Precipitate **insoluble**	Cu^{2+}
Dirty green precipitate	Precipitate **insoluble**	Fe^{2+}
Rusty brown precipitate	Precipitate **insoluble**	Fe^{3+}
No precipitate – **NH$_3$(g)** evolved on warming	—	NH_4^+

Table 23.3 Effect of dilute sodium hydroxide solution

Explanations

The metal cations all form **insoluble hydroxides** with sodium hydroxide solution:

$$M^{n+}(aq) + nOH^-(aq) \longrightarrow M(OH)n(s)$$

e.g.
$$Cu^{2+}(aq) + 2OH^-(aq) \longrightarrow Cu(OH)_2(s)$$

Ca(OH)$_2$, Cu(OH)$_2$, Fe(OH)$_2$ and **Fe(OH)$_3$** are **basic** and therefore do not react with excess sodium hydroxide – precipitates remain.

Al(OH)$_3$, Pb(OH)$_2$ and **Zn(OH)$_2$** are **amphoteric**, therefore react with excess sodium hydroxide forming **soluble** salts – precipitates disappear.

The **ammonium** cation does not form an insoluble hydroxide but, on heating, reacts with the hydroxide ion forming ammonia and water:

$$OH^-(aq) + NH_4^+(aq) \longrightarrow H_2O(l) + NH_3(g)$$

3 EFFECT OF DILUTE AMMONIA SOLUTION (AMMONIUM HYDROXIDE)

a Prepare a **solution** of the solid.
b Add dilute ammonia solution **dropwise**. Look for a **precipitate**.
c Add **excess** ammonia solution. Look for the precipitate **redissolving**.

See Table 23.4.

Explanations

Ammonia solution is a **weak** alkali. Metal cations, except Ca^{2+}, form **insoluble hydroxides** with ammonia solution as with sodium hydroxide solution (see above).

Effect of ammonia solution		Cation
Dropwise	**Excess**	
No precipitate	—	Ca^{2+}
White precipitate	Precipitate **insoluble**	Al^{3+}, Pb^{2+}
White precipitate	Precipitate **soluble** – dissolves forming colourless solution	Zn^{2+}
Blue precipitate	Precipitate **soluble** – dissolves forming deep blue solution	Cu^{2+}
Dirty green precipitate	Precipitate **insoluble**	Fe^{2+}
Rusty brown precipitate	Precipitate **insoluble**	Fe^{3+}

Table 23.4 Effect of dilute ammonia solution (ammonium hydroxide)

The amphoteric hydroxides, **Al(OH)$_3$** and **Pb(OH)$_2$**, and the basic hydroxides, **Fe(OH)$_2$** and **Fe(OH)$_3$**, do not react with excess ammonia solution – precipitates remain.

However, **Zn(OH)$_2$** and **Cu(OH)$_2$** do react with excess ammonia solution forming the **complex soluble hydroxides** tetra-ammine zinc hydroxide, $[Zn(NH_3)_4](OH)_2$, and tetra-ammine copper(II) hydroxide, $[Cu(NH_3)_4](OH)_2$ – precipitates disappear.

4 FURTHER CONFIRMATORY TESTS

Combining Tests 2 and 3, the only cations which cannot be distinguished are **Al^{3+}** and **Pb^{2+}**. Use the tests in Table 23.5 to distinguish Al^{3+} and Pb^{2+}, and to confirm other ions. Carry out all tests on **solutions** of the compound.

Test	Observation	Cation
Add potassium iodide solution	**Bright yellow** precipitate (PbI_2) **No** precipitate	Pb^{2+} Al^{3+}
Add dilute hydrochloric acid	**White** precipitate, dissolves on heating, re-precipitates on cooling ($PbCl_2$) **No** precipitate	Pb^{2+} Al^{3+}
Add potassium hexacyanoferrate(III) solution	**Dark blue** precipitate	Fe^{2+}
Add potassium hexacyanoferrate(II) solution	**Dark blue** precipitate **Red-brown** precipitate **White** precipitate	Fe^{3+} Cu^{2+} Zn^{2+}
Add potassium or ammonium thiocyanate solution	**Deep red** solution	Fe^{3+}

Table 23.5 Further confirmatory tests

Tests to identify anions

1 EFFECT OF HEAT

Heat a sample of the **solid** in a **dry** test tube.

Observation	Anion	Explanation
Carbon dioxide evolved	CO_3^{2-} (or HCO_3^-) – not of K or Na	$CO_3^{2-}(s) \xrightarrow{\text{heat}} O^{2-}(s) + CO_2(g)$
Oxygen evolved	NO_3^- of K or Na	$2NO_3^-(s) \xrightarrow{\text{heat}} 2NO_2^-(s) + O_2(g)$
Oxygen and **nitrogen dioxide** evolved	NO_3^- of Ca and below in reactivity series	$4NO_3^-(s) \xrightarrow{\text{heat}} 2O^{2-}(s) + 4NO_2(g) + O_2(g)$
Sulphur dioxide evolved	SO_3^{2-}	$SO_3^{2-}(s) \xrightarrow{\text{heat}} O^{2-}(s) + SO_2(g)$

Table 23.6 Effect of heat

2 EFFECT OF DILUTE ACID

Add dilute hydrochloric or nitric acid to a sample of the **solid**. Warm if no reaction occurs when cold.

Observation	Anion	Explanation
Carbon dioxide evolved	CO_3^{2-} (or HCO_3^-)	$CO_3^{2-}(s) + 2H^+(aq) \longrightarrow H_2O(l) + CO_2(g)$
Sulphur dioxide evolved	SO_3^{2-}	$SO_3^{2-}(s) + 2H^+(aq) \longrightarrow H_2O(l) + SO_2(g)$

Table 23.7 Effect of dilute acid

3 EFFECT OF CONCENTRATED SULPHURIC ACID

Add concentrated sulphuric acid to a sample of the **solid**. Warm if no reaction occurs when cold.

Observation	Anion	Explanation
Carbon dioxide evolved	CO_3^{2-} (or HCO_3^-)	Carbonic acid is formed, which then decomposes releasing CO_2
Sulphur dioxide evolved	SO_3^{2-}	Sulphurous acid is formed, which then decomposes releasing SO_2
Hydrogen chloride evolved	Cl^-	$2Cl^-(s) + H_2SO_4(l) \longrightarrow SO_4^{2-}(s) + 2HCl(g)$
Hydrogen bromide (colourless) and **brown bromine** vapour evolved	Br^-	$2Br^-(s) + H_2SO_4(l) \longrightarrow SO_4^{2-}(s) + 2HBr(g)$ HBr is then oxidised by conc. H_2SO_4 to Br_2
Black iodine present and **purple iodine** vapour evolved	I^-	$2I^-(s) + H_2SO_4(l) \longrightarrow SO_4^{2-}(s) + 2HI(g)$ HI is immediately oxidised by conc. H_2SO_4 to I_2

Table 23.8 Effect of concentrated sulphuric acid

4 EFFECT OF SILVER NITRATE SOLUTION FOLLOWED BY AQUEOUS AMMONIA

Make a **solution** of the solid in dilute nitric acid. Add a few drops of silver nitrate solution and observe the **precipitate**. Add ammonia solution and observe the **solubility** of the precipitate.

Effect of silver nitrate solution	Anion	Explanation	Effect of aqueous ammonia
White precipitate – turns purple in sunlight	**Cl⁻**	$Ag^+(aq) + Cl^-(aq) \longrightarrow AgCl(s)$ white	Precipitate **soluble**
Cream precipitate – turns yellow-green in sunlight	**Br⁻**	$Ag^+(aq) + Br^-(aq) \longrightarrow AgBr(s)$ cream	Precipitate **slightly soluble**
Pale yellow precipitate	**I⁻**	$Ag^+(aq) + I^-(aq) \longrightarrow AgI(s)$ pale yellow	Precipitate **insoluble**

Table 23.9 Effect of silver nitrate solution and aqueous ammonia

5 EFFECT OF BARIUM NITRATE OR CHLORIDE SOLUTION FOLLOWED BY DILUTE ACID

Make a **solution** of the solid. Add barium nitrate or chloride solution and observe the **precipitate**. Add dilute hydrochloric or nitric acid, warm if necessary and observe the **solubility** of the precipitate.

Effect of $Ba(NO_3)_2(aq)$ or $BaCl_2(aq)$	Effect of HCl(aq) or HNO_3(aq)	Anion	Explanation
White precipitate	Precipitate **insoluble**	SO_4^{2-}	$Ba^{2+}(aq) + SO_4^{2-}(aq) \longrightarrow BaSO_4(s)$ white $BaSO_4$ does not react with dilute acids
White precipitate	Precipitate **soluble** – CO_2 evolved	CO_3^{2-}	$Ba^{2+}(aq) + CO_3^{2-}(aq) \longrightarrow BaCO_3(s)$ white $BaCO_3(s) + 2HCl(aq) \longrightarrow$ $BaCl_2(aq) + H_2O(l) + CO_2(g)$
White precipitate	Precipitate **soluble** – SO_2 evolved on warming	SO_3^{2-}	$Ba^{2+}(aq) + SO_3^{2-}(aq) \longrightarrow BaSO_3(s)$ white $BaSO_3(s) + 2HCl(aq) \longrightarrow$ $BaCl_2(aq) + H_2O(l) + SO_2(g)$

Table 23.10 Effect of barium nitrate or chloride solution and dilute acid

6 FURTHER TESTS FOR NITRATE (NO_3^-) IONS

Test	Observations
Add concentrated sulphuric acid and copper turnings to **solid**. Warm gently	**Blue** solution forms and **nitrogen dioxide** is evolved
Brown ring test: make a **solution** of the solid. Add saturated iron(II) sulphate solution and mix. Add concentrated sulphuric acid down side of test tube	Sulphuric acid sinks, **brown ring** forms between the two liquid layers

Table 23.11 Two further tests for nitrate ions

24 An introduction to organic chemistry

Organic chemistry is the chemistry of compounds which contain **carbon (organic compounds)**. Most also contain **hydrogen** and many contain **oxygen** and/or other elements.

Organic compounds

THE STRUCTURE OF ORGANIC COMPOUNDS

Carbon has a valency of **four** and normally forms **covalent** bonds.

Example

Methane, CH$_4$

$$\cdot C \cdot + 4\overset{\times}{H} \longrightarrow H \overset{\bullet}{\underset{\times}{\times}} C \overset{\times}{\underset{\bullet}{\bullet}} H \quad or \quad H - \overset{\overset{\displaystyle H}{|}}{\underset{\underset{\displaystyle H}{|}}{C}} - H$$

1 carbon atom   4 hydrogen atoms   1 methane molecule

NB The shape of the methane molecule is **tetrahedral**:

(projects backwards)   Bond angle = 109.5°   H (projects forwards)

As a result of carbon having **four valency electrons**, carbon atoms can join with each other in an almost unlimited way to form the following:

1 Straight or **branched chains** of different lengths,

e.g.

$$H-\underset{\underset{H}{|}}{\overset{\overset{H}{|}}{C}}-\underset{\underset{H}{|}}{\overset{\overset{H}{|}}{C}}-\underset{\underset{H}{|}}{\overset{\overset{H}{|}}{C}}-\underset{\underset{H}{|}}{\overset{\overset{H}{|}}{C}}-H$$

$$H-\underset{\underset{H}{|}}{\overset{\overset{H}{|}}{C}}-\underset{\underset{H}{|}}{\overset{\overset{H}{|}}{C}}-\underset{\underset{\underset{\underset{\underset{H}{|}}{C}}{|}}{\overset{\overset{H}{|}}{\underset{|}{H-C-H}}}}{C}}-\underset{\underset{H}{|}}{\overset{\overset{H}{|}}{C}}-\underset{\underset{H}{|}}{\overset{\overset{H}{|}}{C}}-H$$

2 Rings of different sizes,

e.g.

3 Single, **double** or **triple** bonds,

e.g.

FORMULAE OF ORGANIC COMPOUNDS

Formulae of organic compounds may be written in **four** ways:

- **The molecular formula**: this shows the **actual number** of atoms of each element in one molecule of the compound.
- **The empirical formula**: this shows the simplest **ratio** of atoms of each element present in the compound.
- **The full structural formula**: this shows, in two-dimensional **diagrammatic** form, how the atoms are **arranged** in one molecule.
- **The shortened structural formula**: this shows the **sequence** and **arrangement** of atoms in one molecule in such a way that the nature and position of attachment of each **functional group** is shown without actually drawing the molecule.

Example

Butanoic acid

- Molecular formula: $C_4H_8O_2$
- Empirical formula: C_2H_4O
- Full structural formula:

- Shortened structural formula: $CH_3CH_2CH_2COOH$ (which may be shortened further to C_3H_7COOH).

NB
- Each **carbon** atom forms **four** bonds with other atoms.
- Each **nitrogen** atom forms **three** bonds with other atoms.
- Each **oxygen** atom forms **two** bonds with other atoms.
- Each **hydrogen** atom forms **one** bond with another atom.

ISOMERISM

Isomerism *is the occurrence of two or more organic compounds with the same molecular formula but different structural formulae. The compounds are called* **isomers**.

Example C_5H_{12}

This has **three** isomers:

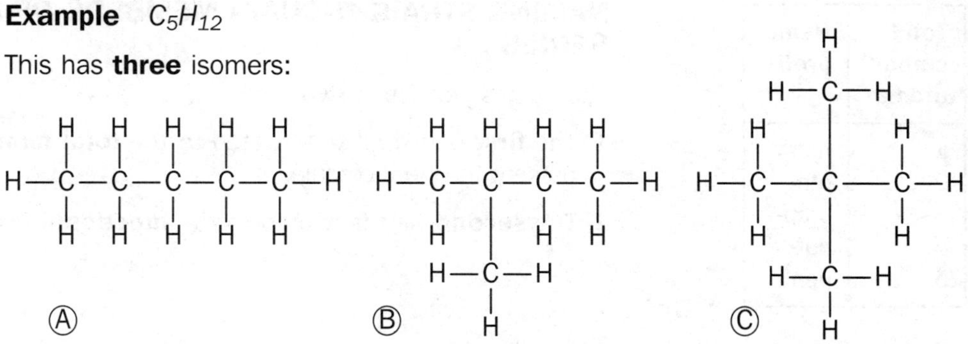

Ⓐ

$CH_3CH_2CH_2CH_2CH_3$

Ⓑ

$CH_3CH(CH_3)CH_2CH_3$

Ⓒ

$CH_3C(CH_3)_2CH_3$

NB The structures drawn below are **not** further isomers:

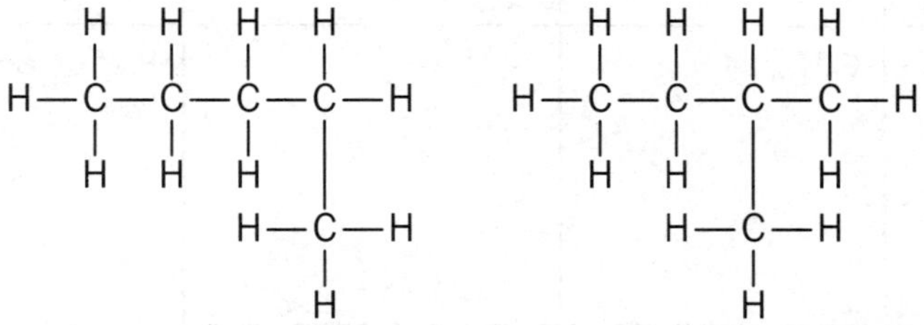

$CH_3CH_2CH_2CH_2CH_3$

This is the same as Ⓐ

$CH_3CH_2CH(CH_3)CH_3$

This is a mirror image of Ⓑ

To check if structures are isomers, write their shortened structural formulae as shown above. If two formulae are the **same**, reading forwards or backwards, then the structures are the same isomer.

FUNCTIONAL GROUPS

Each organic compound is made up of **two** parts:

- The **hydrocarbon part** composed of carbon and hydrogen atoms.
- The **functional group or groups** comprising another atom or group of atoms. The reactions of the functional group(s) determine the **chemical properties** of the compound.

Homologous series

Organic compounds are **classified** into distinct groups called **homologous series**. Homologous series have the following characteristics:

1 All members contain the same **functional group**.

2 All members have the same **general formula**.

3 All members show similar **chemical properties**. Reactivity **decreases** as number of carbon atoms per molecule (molecular mass) **increases**.

4 **Physical properties** show **gradation** along a series. As the number of carbon atoms per molecule **increases**, melting point, boiling point and density **increase**, solubility in water **decreases**.

5 Each member differs in molecular formula from the next by **CH₂**, or in relative molecular mass by **14**.

6 All members may be **prepared** by similar methods.

Total carbon atoms	Name prefix
1	meth-
2	eth-
3	prop-
4	but-
5	pent-

NAMING STRAIGHT-CHAIN MEMBERS OF A HOMOLOGOUS SERIES

The names consist of **two** parts:

1 The **first** part (prefix) is related to the **total number of carbon atoms** present in one molecule.

2 The **second** part is related to the **functional group** present (see Table 24.1).

Homologous series	General formula	Functional group	Naming	Example containing two carbon atoms
Alkanes	C_nH_{2n+2}	—	Prefix + **ane**	Ethane, C_2H_6 H H | | H—C—C—H | | H H
Alkenes	C_nH_{2n}	Carbon-carbon double bond: C=C	Prefix + **ene**	Ethene, C_2H_4 H H \ / C=C / \ H H
Alkynes	C_nH_{2n-2}	Carbon-carbon triple bond: —C≡C—	Prefix + **yne**	Ethyne, C_2H_2 H—C≡C—H
Alcohols	$C_nH_{2n+1}OH$	Hydroxyl group, —OH —O—H	Prefix + **anol**	Ethanol, C_2H_5OH H H | | H—C—C—O—H | | H H
Alkanoic (fatty or carboxylic) acids	$C_nH_{2n+1}COOH$	Carboxyl group, —COOH O ‖ —C \ O—H	Prefix + **anoic acid**	Ethanoic acid, CH_3COOH H O | ‖ H—C—C | \ H O—H
Amines	$C_nH_{2n+1}NH_2$	Amino group, —NH_2 H | —N | H	Prefix + **ylamine**	Ethylamine, $C_2H_5NH_2$ H H H | | | H—C—C—N | | \ H H H

NB C_nH_{2n+1} is called the **alkyl** group, e.g. C_2H_5 = ethyl, C_4H_9 = butyl. **R** may be used to represent the alkyl group.

Table 24.1 The main homologous series

25 The hydrocarbons

Hydrocarbons are organic compounds containing **carbon** and **hydrogen** atoms only. They include the alkanes, alkenes, alkynes and aromatics (ringed hydrocarbons).

Natural sources of hydrocarbons

1 **Natural gas** consists mainly of methane (CH_4) with small amounts of ethane (C_2H_6), propane (C_3H_8) and butane (C_4H_{10}).
2 **Petroleum (crude oil)** is a complex mixture of hydrocarbons; the simplest gaseous ones, e.g. ethane, are dissolved in the liquid.

FRACTIONAL DISTILLATION OF CRUDE OIL

Crude oil is separated into different **fractions** by fractional distillation:

1 The crude oil is treated to remove **sulphur**, this reduces pollution.
2 The crude oil is then heated and the vapours forced up a **fractionating tower** containing trays and bubble caps. The temperature of the trays **decreases upwards** – the lower the boiling point, the further the vapour will rise before condensing. Different fractions are, therefore, collected at different **heights** and **temperatures**. Each fraction is a **mixture** of hydrocarbons with boiling points within a specific range (see Fig. 25.1).

CRACKING

This involves **breaking up** larger hydrocarbon molecules into smaller ones. It is important since distillation of crude oil produces excess larger hydrocarbons and insufficient smaller, more useful hydrocarbons, to meet modern demands.

- **Thermal cracking** uses heat.
- **Catalytic cracking** uses heat plus a catalyst.

Cracking always results in the formation of at least one **alkene** and is, therefore, a major source of alkenes, e.g.

C_5H_{12} pentane C_3H_8 propane C_2H_4 ethene

These smaller hydrocarbons, especially the alkenes, form the foundation of the **petrochemical industry** from which thousands of other compounds are manufactured (see Uses of alkenes, p. 133).

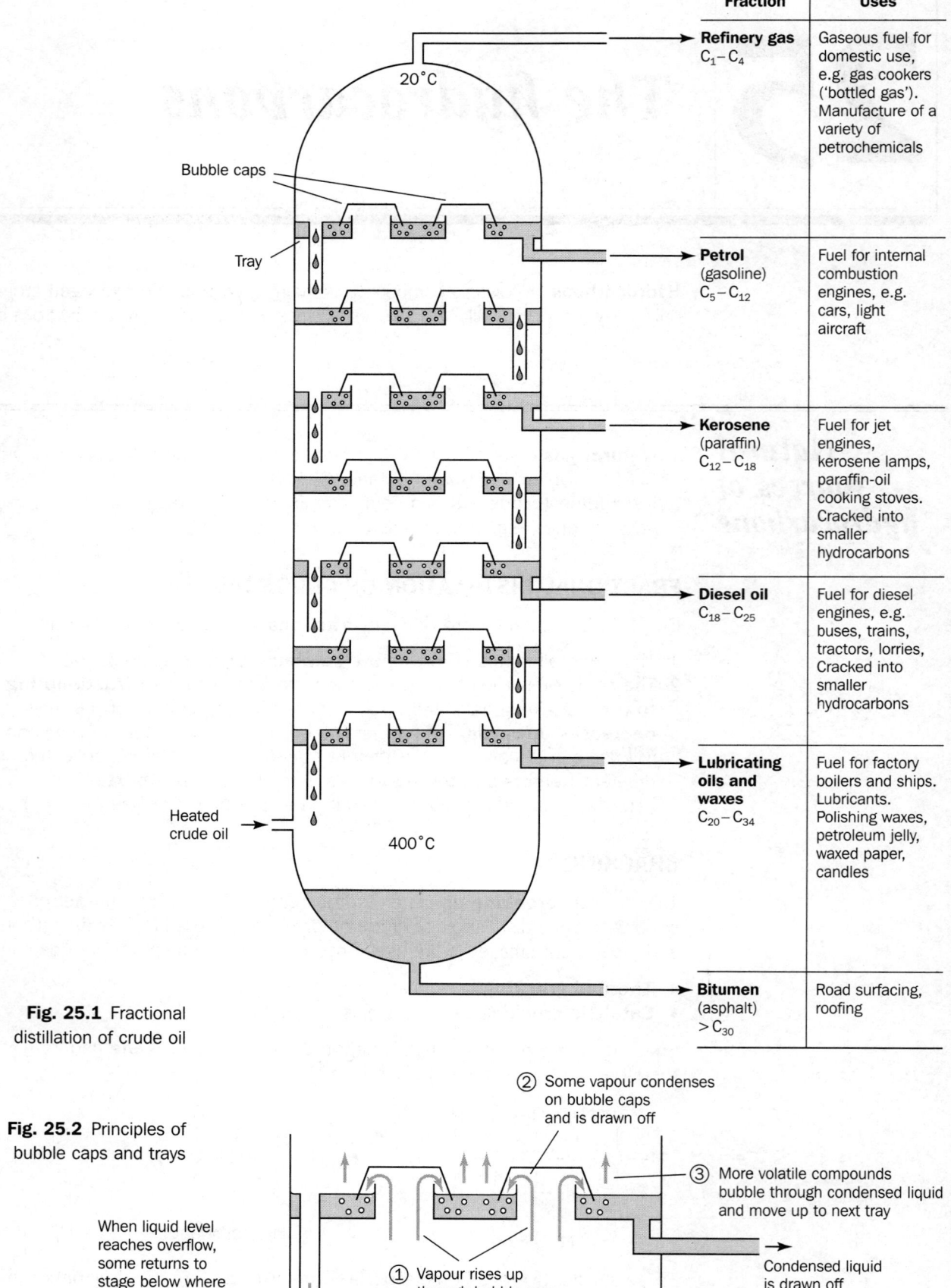

Fraction	Uses
Refinery gas $C_1 - C_4$	Gaseous fuel for domestic use, e.g. gas cookers ('bottled gas'). Manufacture of a variety of petrochemicals
Petrol (gasoline) $C_5 - C_{12}$	Fuel for internal combustion engines, e.g. cars, light aircraft
Kerosene (paraffin) $C_{12} - C_{18}$	Fuel for jet engines, kerosene lamps, paraffin-oil cooking stoves. Cracked into smaller hydrocarbons
Diesel oil $C_{18} - C_{25}$	Fuel for diesel engines, e.g. buses, trains, tractors, lorries. Cracked into smaller hydrocarbons
Lubricating oils and waxes $C_{20} - C_{34}$	Fuel for factory boilers and ships. Lubricants. Polishing waxes, petroleum jelly, waxed paper, candles
Bitumen (asphalt) $> C_{30}$	Road surfacing, roofing

20°C

Bubble caps

Tray

Heated crude oil

400°C

Fig. 25.1 Fractional distillation of crude oil

Fig. 25.2 Principles of bubble caps and trays

When liquid level reaches overflow, some returns to stage below where its temperature increases and it revaporises

① Vapour rises up through bubble caps

② Some vapour condenses on bubble caps and is drawn off

③ More volatile compounds bubble through condensed liquid and move up to next tray

Condensed liquid is drawn off

The alkanes: C_nH_{2n+2}

Alkanes are **saturated** hydrocarbons. Saturated compounds contain only **single** covalent bonds between carbon atoms. Straight and branched-chain isomers occur in alkanes with **four or more** carbon atoms.

At room temperature, C_1 to C_4 are **gases**, C_5 to C_{17} are **liquids** and C_{18} onwards are **solids**.

Formula	Structural formula	Name
CH_4	H | H—C—H | H	Methane
C_2H_6	H H | | H—C—C—H | | H H	Ethane
C_3H_8	H H H | | | H—C—C—C—H | | | H H H	Propane
C_4H_{10} – two isomers	H H H H | | | | H—C—C—C—C—H | | | | H H H H	Butane
	H H H | | | H—C—C—C—H | | | H H H | H—C—H | H	2-Methylpropane

Table 25.1 The first four alkanes

REACTIONS OF ALKANES

Because alkanes are **saturated**, they are relatively **unreactive**.

1 Alkanes burn in air or oxygen
Alkanes burn with **clean, blue, non-smoky** flames forming **carbon dioxide, water** and **heat**,

e.g. $CH_4(g) + 2O_2(g) \longrightarrow CO_2(g) + 2H_2O(g) \quad \Delta H - ve$

2 Alkanes undergo substitution reactions
In substitution reactions, **hydrogen** atoms in alkane molecules are **replaced** by other atoms, e.g. halogen atoms.

Example

Methane + chlorine

No reaction occurs in the dark; in bright light the reaction is **explosive**; in diffused light **substitution** occurs in stages, **one** chlorine atom at a time:

$$CH_4(g) + Cl_2(g) \longrightarrow \quad CH_3Cl(g) \quad + HCl(g)$$
$$\text{chloromethane}$$

$$CH_3Cl(g) + Cl_2(g) \longrightarrow \quad CH_2Cl_2(l) \quad + HCl(g)$$
$$\text{dichloromethane}$$

$$CH_2Cl_2(l) + Cl_2(g) \longrightarrow \quad CHCl_3(l) \quad + HCl(g)$$
$$\text{trichloromethane}$$

$$CHCl_3(l) + Cl_2(g) \longrightarrow \quad CCl_4(l) \quad + HCl(g)$$
$$\text{tetrachloromethane}$$

A similar substitution occurs with bromine and with other alkanes.

3 Alkanes can be cracked (see p. 129).

USES OF ALKANES

1 Alkanes are used as **fuels** as they burn readily in air, the reactions are highly **exothermic** and the flames are **clean** – do not produce polluting smoke. For example, natural gas is mainly methane, bottled gas is a mixture of propane and butane, petrol (gasoline) is a mixture mainly of alkanes.

2 Alkanes are used in the **manufacture of other chemicals**, e.g. methane is used to make hydrogen for ammonia production.

3 Larger alkanes are used as **solvents** as they readily dissolve many organic compounds, e.g. petroleum jelly is oil dissolved in paraffin wax, a solid alkane.

The alkenes C_nH_{2n}

Alkenes are **unsaturated** hydrocarbons. Unsaturated compounds contain **double** and/or **triple** covalent bonds between two or more carbon atoms. Alkenes have **one double bond** between one pair of carbon atoms. Straight and branched-chain isomers occur in alkenes with **four or more** carbon atoms. Also the **position** of the double bond can vary to produce further isomers. (See Table 25.2.)

REACTIONS OF ALKENES

Alkenes are **reactive** because one bond of the double bond is weaker and is easily broken:

1 **Alkenes burn in air or oxygen**
   Alkenes burn with **smoky yellow** flames which contain unburnt carbon due to the high proportion of carbon in alkene molecules.
   Carbon dioxide, **water** and **heat** are formed,

e.g. $\quad C_2H_4(g) + 3O_2(g) \longrightarrow 2CO_2(g) + 2H_2O(g) \quad \Delta H - \text{ve}$

Formula	Structural formula	Name
C_2H_4	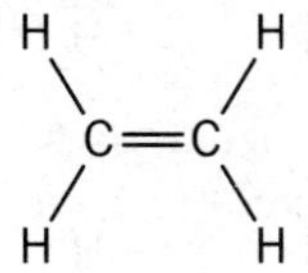	Ethene
C_3H_6	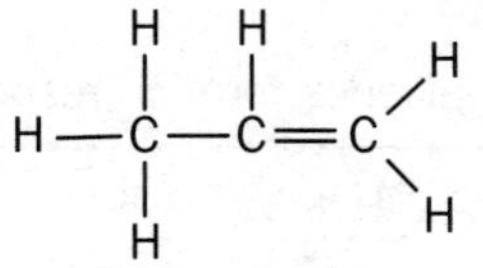	Propene
C_4H_8 – three isomers	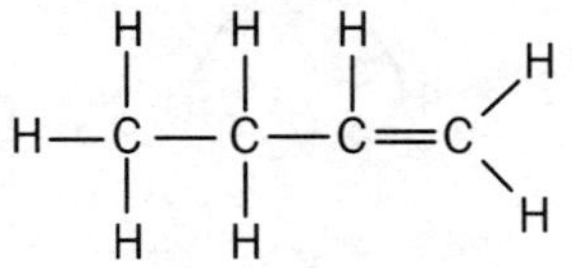	But-1-ene
	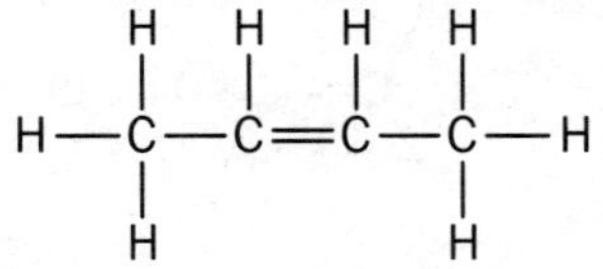	But-2-ene
	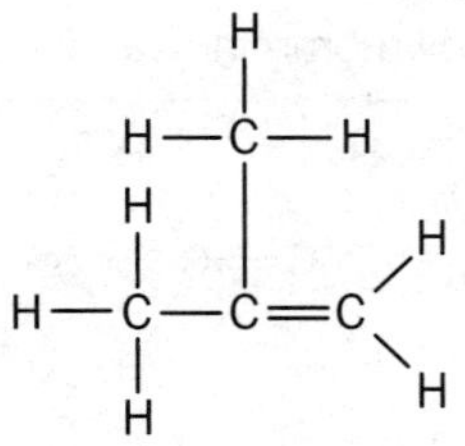	2-Methylpropene

Table 25.2 The first three alkenes

2 Alkenes undergo addition reactions

In addition reactions, **two** molecules react to form **one** molecule:

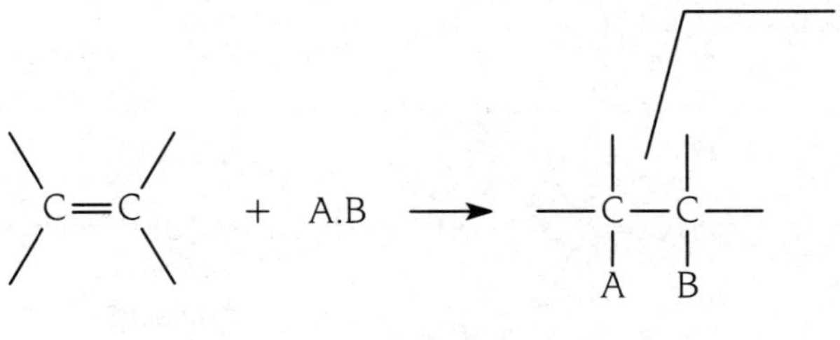

The double bond is converted into a single bond, forming a **saturated** compound. A.B is added across the original double bond

See Table 25.3 for addition reactions of ethene.

Table 25.4 shows tests to distinguish an alkane and an alkene.

USES OF ALKENES

Alkenes are used in the **manufacture** of many other chemicals owing to their readiness to undergo **addition** reactions, e.g. to manufacture ethanol and other alcohols, antifreeze (ethane-1, 2-diol), solvents, detergents, plastics, synthetic rubbers, paints, many medicines.

Reaction	Equation and conditions
Hydrogenation – addition of **hydrogen**	 $CH_2=CH_2$ (g) + H_2(g) $\xrightarrow[\text{200 °C}]{\text{Ni or Pt catalyst}}$ ethane (g) (Other alkenes form corresponding **alkanes**)
Halogenation – addition of **halogens**, e.g. chlorine gas, bromine vapour, or a solution of bromine in water or tetrachloromethane	$CH_2=CH_2$ (g) + Cl_2(g) $\longrightarrow$ 1, 2-dichloroethane (l) $CH_2=CH_2$ (g) + Br_2(g) $\longrightarrow$ 1, 2-dibromoethane (l) Red-brown bromine vapour or bromine solution is rapidly decolorised
Hydration – addition water as **steam**	$CH_2=CH_2$ (g) + H_2O(g) (H.OH) $\xrightarrow[\text{300 °C, 60 atm}]{\substack{H_3PO_4 \text{ in}\\ \text{silica catalyst}}}$ ethanol (l) (Other alkenes form corresponding **alcohols**)
Addition of **hydrogen halides**, e.g. hydrogen chloride, hydrogen bromide, hydrogen iodide	$CH_2=CH_2$ (g) + HI(g) $\longrightarrow$ iodoethane (l)
With **acidified potassium manganate(VII)** solution	$CH_2=CH_2$ (g) + H_2O(l) + [O] $\longrightarrow$ ethane -1, 2-diol (ethylene glycol) (l) from oxidising agent Acidified potassium manganate(VII) solution is rapidly decolorised
Polymerisation	When ethene is heated to 200 °C at high pressure (2000 atm), or subjected to low pressure (25 atm) and a suitable catalyst, polyethene is formed (see Table 27.2, p. 145) (Other alkenes form corresponding **polyalkenes**)

Table 25.3 Addition reactions of ethene

Test	Observations	
	Alkane	**Alkene**
Add bromine vapour	**Slow** decolorisation of bromine in light – slow substitution occurs	**Rapid** decolorisation of bromine in dark or light – rapid addition occurs
Add a solution of bromine in water or tetrachloromethane	**No** colour change – no reaction occurs	**Rapid** decolorisation of bromine solution – rapid addition occurs
Add acidified potassium manganate(VII) solution	**No** colour change – no reaction occurs	**Rapid** decolorisation of acidified potassium manganate(VII) solution – rapid addition occurs

Table 25.4 Distinguishing between an alkane and an alkene

26 Alcohols, alkanoic acids and esters

Alcohols: $C_nH_{2n+1}OH$

Straight and branched-chain isomers occur in alcohols with **four or more** carbon atoms. Also the **position** of the functional group can vary to produce further isomers.

The **properties** of alcohols are determined by the **hydroxyl (–OH) group**. Owing to the presence of this group, alcohols (especially the smaller ones) are **soluble** in water and are **less volatile** (have higher boiling points) than their corresponding alkanes, i.e. at room temperature C_1 to C_{12} are **liquids**, the rest are **solids**.

Formula	Structural formula	Name
CH_3OH	H | H—C—O—H | H	Methanol
C_2H_5OH	H H | | H—C—C—O—H | | H H	Ethanol
C_3H_7OH	H H H | | | H—C—C—C—O—H | | | H H H	Propanol
C_4H_9OH	H H H H | | | | H—C—C—C—C—O—H | | | | H H H H	Butanol

Table 26.1 The first four alcohols (excluding isomers)

PREPARATION OF ETHANOL

Fermentation of carbohydrates

Ethanol is made by adding **yeast cells** to **glucose** solution. The enzyme **zymase**, produced by the yeast cells, breaks down the glucose **anaerobically** (without oxygen) to produce **ethanol, carbon dioxide** and **energy**

for the cells:

$$C_6H_{12}O_6(aq) \xrightarrow[\text{yeast cells}]{\text{zymase in}} 2C_2H_5OH(aq) + 2CO_2(g) + \text{energy}$$

When the concentration of ethanol reaches about 14%, it poisons the yeast cells. 96% ethanol is obtained by **fractional distillation**.

Wine is made by adding yeast to **grapes**. Zymase from the yeast cells ferments the glucose in the grape juice. Air must not come into contact with the wine, since certain bacteria **oxidise** ethanol to **ethanoic acid (vinegar)** which causes the wine to become sour:

$$C_2H_5OH(aq) + O_2(g) \xrightarrow[\text{bacteria}]{\text{aerobic}} CH_3COOH(aq) + H_2O(l)$$
$$\text{ethanoic acid}$$

Rum is made by adding yeast to **molasses**. The yeast cells produce the enzyme **invertase** which **digests** the sucrose in the molasses into glucose and fructose. The yeast then **ferments** the glucose and fructose, and the mixture is **distilled**.

Hydration of ethene – the industrial preparation

Ethene and **steam** are passed over a catalyst of phosphoric acid absorbed on silica at 60 atm and 300 °C (see Table 25.3, p. 134).

REACTIONS OF ETHANOL

1 **Ethanol burns in air or oxygen**.
   Ethanol burns with a **clean blue** flame forming **carbon dioxide**, **water** and **heat**:

$$C_2H_5OH(l) + 3O_2(g) \longrightarrow 2CO_2(g) + 3H_2O(g) \qquad \Delta H \text{ } -\text{ve}$$

2 **Ethanol reacts with sodium**.
   Sodium ethoxide and **hydrogen** are formed:

$$2C_2H_5OH(l) + 2Na(s) \longrightarrow 2C_2H_5ONa(\text{alc. soln.}) + H_2(g)$$
$$\text{sodium ethoxide}$$

3 **Ethanol can be dehydrated**.
   Ethanol is dehydrated to **ethene** by heating it to about 170 °C in the presence of concentrated sulphuric acid, or by passing ethanol vapour over heated aluminium oxide:

$$C_2H_5OH(l) \xrightarrow[\text{170 °C}]{\text{conc. H}_2\text{SO}_4} C_2H_4(g) + H_2O(l)$$

4 **Ethanol can be oxidised**.
   Ethanol is oxidised to **ethanoic acid** by heating with acidified potassium manganate(VII) solution, or acidified potassium dichromate(VI) solution:

$$C_2H_5OH(aq) + \underset{\substack{\text{from} \\ \text{oxidising} \\ \text{agent}}}{2[O]} \longrightarrow CH_3COOH(aq) + H_2O(l)$$

Potassium dichromate(VI) crystals moistened with sulphuric acid are used in the **breathalyser test** to detect drunken drivers. If the crystals turn from **orange** to **dark green** when breathed over, as a result of being reduced, the driver's breath contains ethanol vapour.

5 Ethanol reacts with alkanoic acids.
An **ester** and **water** are formed (see p. 140).

USES OF ALCOHOLS

1 Alcohols are used as **fuels** as they burn readily in air. The reactions are fairly **exothermic** and the flames are **clean**, e.g. methylated spirits is mainly ethanol, gasohol is petrol with ethanol added.
2 Alcohols are used as **solvents** for many industrial and pharmaceutical chemicals, e.g. varnishes, perfumes.
3 Ethanol is used in producing **alcoholic beverages**.

ALCOHOL ABUSE

Alcohol is a **depressant** of the central nervous system. People who repeatedly drink alcoholic beverages in excess of normal social drinking levels run the risk of becoming **alcoholics**. Alcoholics are dependent on alcohol, **alcoholism** being classed as a **disease**.

Short-term effects of alcohol abuse include:

- Impaired mental functioning
- Impaired muscular skills, co-ordination and judgement
- Slowed reflexes
- Slurred speech
- Memory loss
- Loss of consciousness.

Long-term effects of alcohol abuse include:

- Aggravation of hypertension (high blood pressure)
- Cirrhosis (hardening) of the liver
- Intestinal, skeletal muscle and nervous disorders
- Delirium tremens (DTs).

Alkanoic acids: $C_nH_{2n+1}COOH$

Straight and branched-chain isomers occur in alkanoic acids with **four or more** carbon atoms. Also the **position** of the functional group can vary to produce further isomers.

The **properties** of alkanoic acids are due to the presence of the **carboxyl (–COOH) group**. Like alcohols, alkanoic acids are **soluble** in water, however, they are **less volatile** than their corresponding alcohols.

Formula	Structural formula	Name
HCOOH	$$H-C\begin{matrix}O\\ \\O-H\end{matrix}$$	Methanoic acid
CH_3COOH	$$H-\overset{H}{\underset{H}{C}}-C\begin{matrix}O\\ \\O-H\end{matrix}$$	Ethanoic acid

Table 26.2 The first four alkanoic acids (excluding isomers)

(continued)

Formula	Structural formula	Name
C_2H_5COOH		Propanoic acid
C_3H_7COOH		Butanoic acid

Table 26.2 *Continued*

LABORATORY PREPARATION OF ETHANOIC ACID

Ethanoic acid is prepared in the laboratory by the **oxidation of ethanol** (see p. 137). Ethanol is mixed with acidified potassium dichromate(VI) solution and heated under **reflux** (see Fig 26.1, p. 141). The mixture is then **fractionally distilled** to obtain the ethanoic acid.

REACTIONS OF ANHYDROUS ETHANOIC ACID

1 **Anhydrous ethanoic acid burns in air or oxygen**.
   Ethanoic acid burns forming **carbon dioxide**, **water** and **heat**:

$$CH_3COOH(l) + 2O_2(g) \longrightarrow 2CO_2(g) + 2H_2O(g) \qquad \Delta H \text{ } -ve$$

2 **Anhydrous ethanoic acid reacts with alcohols**.
   An **ester** and **water** are formed (see p. 140).

REACTIONS OF AQUEOUS ETHANOIC ACID

Ethanoic acid is **partially** ionised in water, therefore it is a **weak acid** which reacts in the same way as other acids:

$$CH_3COOH(aq) \rightleftharpoons CH_3COO^-(aq) + H^+(aq)$$
$$\text{ethanoate ion}$$

Salts of ethanoic acid are called **ethanoates**.

1 **Aqueous ethanoic acid reacts with reactive metals**.
   A **salt** and **hydrogen** are produced,

   e.g.

$$2CH_3COOH(aq) + Mg(s) \longrightarrow (CH_3COO)_2Mg(aq) + H_2(g)$$
$$\text{magnesium ethanoate}$$

2 **Aqueous ethanoic acid reacts with oxides and hydroxides of metals**.
   A **salt** and **water** are produced,

   e.g.

$$2CH_3COOH(aq) + CuO(s) \longrightarrow (CH_3COO)_2Cu(aq) + H_2O(l)$$
$$\text{copper(II) ethanoate}$$

3 Aqueous ethanoic acid reacts with carbonates and hydrogen carbonates.

A **salt**, **water** and **carbon dioxide** are produced,

e.g.

$$2CH_3COOH(aq) + Na_2CO_3(aq) \longrightarrow 2CH_3COONa(aq) + H_2O(l) + CO_2(g)$$
$$\text{sodium ethanoate}$$

Esters: $C_nH_{2n+1}COOC_xH_{2x+1}$

When heated in the presence of concentrated sulphuric acid, anhydrous alkanoic acids react with alcohols forming an **ester** and **water**, a reaction called **esterification**:

$$\textbf{alkanoic acid} + \textbf{alcohol} \rightleftharpoons \textbf{ester} + \textbf{water}$$

e.g.

ethanoic acid   ethanol   ethyl ethanoate

Esters are sweet-smelling, oily liquids found naturally in fruits and flowers. Vegetable **oils** and animal **fats** are esters of **long-chain acids** and the alcohol **glycerol** (propane-1,2,3-triol). The functional group present in esters is the ester group, i.e.

or $— COO—$.

NAMES AND FORMULAE OF ESTERS

Esters are named after the **acid** and **alcohol** from which they are derived:

Formula: **acid** part first; **alcohol** part second

Name: **alcohol** part first; **acid** part second

Example

$$\textbf{CH}_3\textbf{COOH}(l) + \textbf{C}_3\textbf{H}_7\textbf{OH}(l) \rightleftharpoons \textbf{CH}_3\textbf{CO} \cdot \textbf{OC}_3\textbf{H}_7(l) + \textbf{H}_2\textbf{O}(l)$$

ethanoic acid + **prop**anol $\rightleftharpoons$ **propyl ethanoate** + water

Other examples:

Acid	Alcohol	Formula and name of ester
HCOOH, **methan**oic acid	C_2H_5OH, **ethan**ol	$HCOOC_2H_5$, **ethyl methanoate**

(continued)

Acid	Alcohol	Formula and name of ester
C_3H_7COOH, **butan**oic acid	CH_3OH, **meth**anol	$C_3H_7COOCH_3$, **methyl butanoate**
C_2H_5COOH, **propan**oic acid	C_4H_9OH, **but**anol	$C_2H_5COOC_4H_9$, **butyl propanoate**

LABORATORY PREPARATION OF ETHYL ETHANOATE

Ethyl ethanoate is prepared by heating ethanol, anhydrous ethanoic acid and concentrated sulphuric acid under **reflux**. The mixture is then added to water and the ester floats since it is **immiscible** with water and **less dense**. The ester is separated using a **separating funnel**. The sulphuric acid is added to:

- **catalyse** the reaction,
- **remove the water** formed, thus favouring the forward reaction (see equation, p. 140).

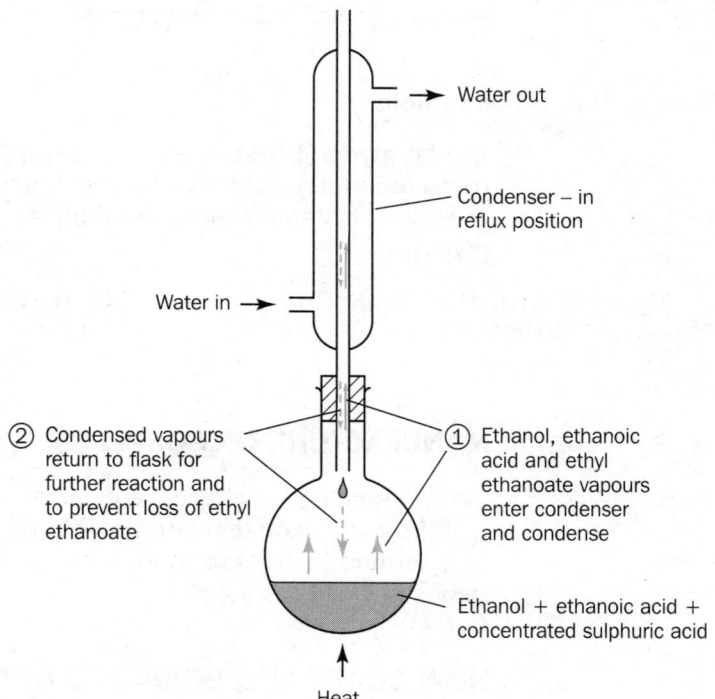

Water out

Condenser – in reflux position

Water in →

② Condensed vapours return to flask for further reaction and to prevent loss of ethyl ethanoate

① Ethanol, ethanoic acid and ethyl ethanoate vapours enter condenser and condense

Ethanol + ethanoic acid + concentrated sulphuric acid

Heat

Fig. 26.1 Laboratory preparation of ethyl ethanoate showing the condenser in the reflux position

HYDROLYSIS OF ESTERS

During **hydrolysis** a substance is broken down by reacting with **water**. Esters are hydrolysed by **boiling** with a dilute aqueous **acid** (H_2SO_4 or HCl) or an aqueous **alkali** (KOH or NaOH):

1 Acid hydrolysis yields the **acid** and **alcohol** that formed the ester,

e.g.

$$CH_3COOC_2H_5(l) + H_2O(l) \underset{}{\overset{H^+(aq)}{\rightleftharpoons}} CH_3COOH(aq) + C_2H_5OH(aq)$$
ethyl ethanoate   ethanoic acid   ethanol

2 Alkaline hydrolysis yields the **salt of the acid** and the **alcohol** that formed the ester. This occurs in two stages:

- The **acid** is formed as in acid hydrolysis,

  e.g.

$$CH_3COOC_2H_5(l) + H_2O(l) \underset{}{\overset{OH^-(aq)}{\rightleftharpoons}} CH_3COOH(aq) + C_2H_5OH(aq)$$

- The acid reacts with excess alkali forming a **salt** and **water**,

  e.g.

$$CH_3COOH(aq) + NaOH(aq) \longrightarrow CH_3COONa(aq) + H_2O(l)$$

Overall reaction:

$$CH_3COOC_2H_5(l) + NaOH(aq) \longrightarrow CH_3COONa(aq) + C_2H_5OH(aq)$$

SAPONIFICATION OF ESTERS

Saponification is the process by which fats and oils are hydrolysed to form **soap** by boiling with **concentrated sodium hydroxide solution**.

Example

The fat **glyceryl tristearate**, $(C_{17}H_{35}COO)_3C_3H_5$, is an ester of **stearic (octadecanoic) acid**, $C_{17}H_{35}COOH$, and **glycerol**, $C_3H_5(OH)_3$. Saponification of glyceryl tristearate yields **sodium stearate**, one form of soap, and **glycerol**:

$$(C_{17}H_{35}COO)_3C_3H_5(l) + 3NaOH(aq) \longrightarrow 3C_{17}H_{35}COONa(aq) + C_3H_5(OH)_3(l)$$
glyceryl tristearate   sodium stearate   glycerol
(a soap)

MANUFACTURE OF SOAPLESS (SYNTHETIC) DETERGENTS

Modern synthetic detergents are made by reacting **hydrocarbons** from petroleum with **concentrated sulphuric acid**. The product is then neutralised with **sodium hydroxide** to form its sodium salt, e.g. sodium dodecanyl sulphate, $C_{12}H_{25}OSO_3Na$.

HOW SOAPY AND SOAPLESS DETERGENTS WORK

Soapy and soapless detergent molecules have a fairly long, **covalent**, **hydrocarbon 'tail'** with an **ionic 'head'**, e.g. $C_{17}H_{35}COO^-Na^+$ or $C_{12}H_{25}OSO_3^-Na^+$.

The 'tail' is **hydrophobic** (water-hating or grease-loving), the 'head' is **hydrophilic** (water-loving).

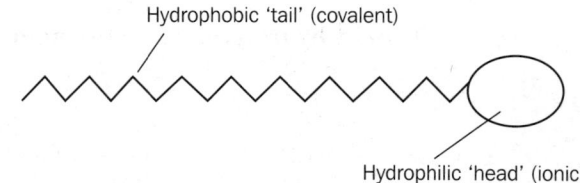

Hydrophobic 'tail' (covalent)

Hydrophilic 'head' (ionic)

Fig. 26.2 A detergent molecule

Detergents work by:

1 **Lowering** the **intermolecular forces** between water molecules, thus lowering the surface tension of water, allowing it to **spread out** and **wet** surfaces more efficiently.
2 **Breaking up** and **dispersing** greasy dirt by the hydrophobic 'tail' entering the grease and the hydrophilic 'head' remaining in the water.

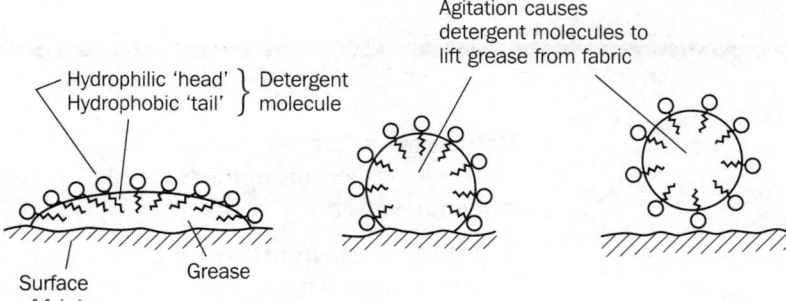

Fig. 26.3 How detergents remove grease from fabric

Soapy detergents	Soapless detergents
Do not lather in hard water – instead they form unpleasant **scum** composed of insoluble calcium and magnesium stearate	Lather in hard water – do not form scum since their calcium and magnesium salts are soluble
Biodegradable, i.e. broken down by bacteria in the environment, therefore do not cause foaming on rivers	Older ones were non-biodegradable – caused foaming on rivers and in sewage systems which prevented oxygen dissolving for aquatic organisms and made sewage treatment difficult. Modern ones are biodegradable
Do not contain phosphates (which pollute)	Contain phosphates which pollute by causing **eutrophication** (see p. 111)
Manufactured from a renewable resource – fats and oils	Manufactured from a non-renewable resource – petroleum

Table 26.3 Soapy and soapless detergents compared

27 Polymers

Polymers are macromolecules formed by linking together **thousands** of small molecules called **monomers**, usually in chains. Polymers are formed by **polymerisation**:

1 **Addition polymerisation** occurs when **unsaturated** monomers are linked to form a **saturated** polymer. The polymer is the **only** product and it has the same empirical formula as the monomer.
2 **Condensation polymerisation** occurs when monomers join with the **elimination** of a small molecule, e.g. water, from between each unit.

The **properties** of polymers depend on:

- The type of **monomer(s)** from which it is formed.
- The type of **linkage** between monomers.

Polymers may be **man-made (synthetic)** or occur **naturally**. Most synthetic polymers are referred to as **plastics**.

Type of polymerisation	Type of polymer	Examples	
		Synthetic	**Natural**
Addition polymerisation	Polyalkenes	Polyethene Polypropene Polystyrene Polyvinyl chloride (PVC)	—
Condensation polymerisation	Polyamides	Nylon	Proteins
	Polyesters	Terylene	(Fats and oils are esters but not polymers)
	Polysaccharides	—	Starch Cellulose Glycogen (animal starch)

Table 27.1 Classification of polymers

Type of polymer	Type of linkage	Structure of monomer(s)	Structure of polymer	Uses
Polyalkene	Alkane	ethene	polyethene	Packaging; manufacture of toys, kitchenware, food containers, buckets, plastic film and bags
		vinyl chloride (monochloroethene)	polyvinyl chloride (PVC)	Insulation for electrical wires; as a building material, e.g. guttering
		other molecules with **C = C**	If X = **CH₃**, polymer is **polypropene**. If X = **C₆H₅**, polymer is **polystyrene**	**Polypropene**: manufacture of ropes, food containers, washing-up bowls **Polystyrene**: packaging, insulation
Polyamide	Amide (peptide) or — CONH —	diacid diamine X and Y are variable hydrocarbon groups	a polyamide e.g. **nylon 66**, where X = C₄H₈, Y = C₆H₁₂	Manufacture of fibres for clothing, ropes, fishing lines
		amino acid amino acid R and R′ are variable hydrocarbon groups which may also contain S and O atoms – there are 20 such groups, therefore 20 different amino acids exist	a protein	Building body cells, hair and nails; making enzymes

Table 27.2 Structure of the common polymers

(continued)

Type of polymer	Type of linkage	Structure of monomer(s)	Structure of polymer	Uses
Polyester	Ester $-\overset{\overset{\displaystyle O}{\|}}{C}-O-$ or $-COO-$	**diacid** and **dialcohol**; $\boxed{X}$ and $\boxed{Y}$ are variable hydrocarbon groups	a polyester; e.g. **Terylene**, where $\boxed{X} = C_6H_4$, $\boxed{Y} = C_2H_4$	Manufacture of fibres for clothing, boat sails, fishing lines
Polysaccharide	Ether $-O-$	**monosaccharide**; e.g. glucose or fructose, $C_6H_{12}O_6$, where $\boxed{X}$ represents $C_6H_{10}O_4$	a polysaccharide	Respired by living organisms after being digested to monosaccharides: $C_6H_{12}O_6(aq) + 6O_2(g) \longrightarrow 6CO_2(g) + 6H_2O(l) + energy$; Stored as food reserves in living organisms, e.g. starch in plants, glycogen in animals

NB $\begin{array}{c} \text{-O-H H-} \end{array}$ represents loss of H_2O from between monomers.

Table 27.2 *Continued*

Advantages	Disadvantages
1 Easily shaped and moulded 2 Inexpensive 3 Light in weight 4 Easily coloured 5 Durable – do not rust, corrode or decay 6 Good thermal and electrical insulators 7 Can be made flexible or rigid 8 Some are very strong	1 Non-biodegradable – contribute to land pollution 2 Produce dense smoke and poisonous gases when burnt – contribute to air pollution 3 Many are flammable – pose fire hazards 4 Difficult to recycle

Table 27.3 Advantages and disadvantages of plastics

Poly-saccharides and proteins

HYDROLYSIS OF POLYSACCHARIDES AND PROTEINS

1 In the laboratory

Polysaccharides and proteins can be hydrolysed to **monosaccharides** and **amino acids**, respectively, by boiling with **dilute acid**.

2 In biological systems

Hydrolysis is achieved by **enzymes** during **digestion**:

$$\text{starch} \atop (C_6H_{10}O_5)_n \xrightarrow{\text{carbohdrases}^*} \text{maltose} \atop {C_{12}H_{22}O_{11} \atop \text{(a disaccharide)}} \xrightarrow{\text{maltase}^*} \text{glucose} \atop C_6H_{12}O_6$$

$$\text{proteins} \xrightarrow{\text{proteases}^*} \text{amino acids}$$

* Digestive enzymes

The **structure** of a protein can be determined by hydrolysing the protein and identifying the amino acids present by **chromatography**.

Property	Glucose	Starch
Solubility in water	**Soluble**	**Insoluble** – forms a colloid
Reaction when heated with Benedict's or Fehling's solution (blue)	**Brick red** precipitate forms – glucose **reduces** the blue copper(II) compound to brick red, insoluble copper(I) oxide	**No** reaction, solution remains blue – starch is not a reducing agent
Reaction with iodine solution	**No** reaction, solution remains brown	**Blue-black** coloration produced

Table 27.4 Differences in the properties of a monomer (glucose) and its polymer (starch)

Sucrose: $C_{12}H_{22}O_{11}$

Sucrose is a **dimer** formed by the **condensation** of the isomers glucose and fructose:

$$C_6H_{12}O_6 + C_6H_{12}O_6 \longrightarrow C_{12}H_{22}O_{11} + H_2O$$
$$\text{glucose} \quad\quad \text{fructose} \quad\quad\quad \text{sucrose}$$

Commercial sources of sucrose are **sugar beet** and **sugar cane**. The monosaccharides, which are produced in **photosynthesis**, are converted to sucrose by these plants and stored in their roots and stems, respectively.

Revision questions

In keeping with the revision nature of the text, the following questions are for revision purposes. They are designed to test mainly **knowledge** and **comprehension**. This should enable you to test yourself and assess your knowledge and understanding by referring to the text for the answers. Most of the questions are not designed specifically to test use of knowledge since this skill must be developed gradually by yourself, with assistance from your teacher, during the two year programme. You cannot readily assess your own ability to use your knowledge without guidance from your teacher.

The questions are mainly of the guided essay-type. The length of individual questions vary, however, since they attempt to give a comprehensive cover of each topic instead of being modelled exactly on the 20 minute questions of Paper 3. A list of relative atomic masses of elements referred to in calculations and answers to questions requiring a numerical solution are given on p. 158.

THE STATES OF MATTER

1 **a** Imagine you are a particle in steam. Describe fully what happens to you as the steam is cooled and changes to water and then to ice.

   b When powdered chalk is suspended in water and observed under the microscope, the chalk particles appear to move in a random way. Explain the cause of this random motion.

   c If a crystal of potassium manganate(VII) is placed at the bottom of a beaker of water and left undisturbed until there is no further change, what would you observe? Explain your observations.

   d A piece of apparatus was set up as in the diagram opposite.

   i Describe what you would observe after the apparatus had been left for about 15 minutes.

   ii Account fully for your observations.

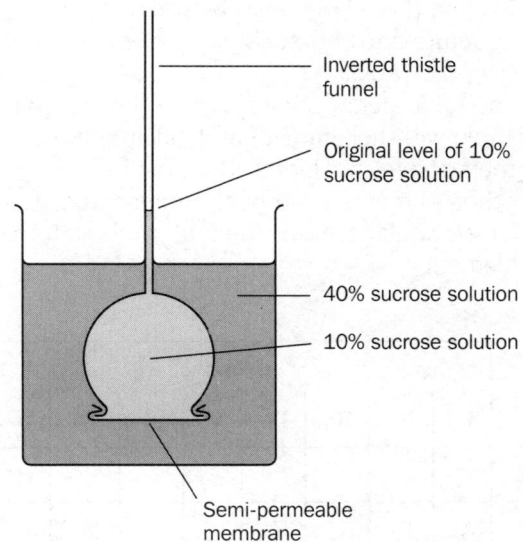

Inverted thistle funnel

Original level of 10% sucrose solution

40% sucrose solution

10% sucrose solution

Semi-permeable membrane

2 **a** Distinguish between an element, a compound and a mixture.

   b Whilst on the beach, you collect a sample of sand and sea water and then return to the laboratory with this sample.

   i Explain how you would separate the sand from the sea water.

   ii Draw a labelled diagram of the apparatus you would use to obtain pure water from the sea water.

   c Explain clearly the **principles** involved in separating:

   i Ethanol and water

   ii Vegetable oil and water

   iii The dyes in a sample of black ink.

ATOMIC STRUCTURE

3 **a** The following symbols refer to atoms of fluorine, calcium and boron:
   $^{19}_{9}F$, $^{40}_{20}Ca$, $^{10}_{5}B$, $^{11}_{5}B$.

   i Show, by diagrams, the structure of a fluorine atom and a calcium atom.

   ii What can you deduce about the two boron atoms?

iii How do you account for the fact that the relative atomic mass of naturally occurring boron is 10.8?

b What is a radioisotope?

c Outline **four** uses of radioisotopes.

THE PERIODIC TABLE

4 a Explain the relationship between atomic structure and the position of elements in the Periodic Table.

b The table below shows part of the Periodic Table with helium (He) and calcium (Ca) in their correct positions. The positions of 12 other elements have been represented by letters which are not the real symbols for the elements.

	Group							
	I	**II**	**III**	**IV**	**V**	**VI**	**VII**	**0**
1								He
2	A	J		D			M	
3	X	L	G			E	Q	
4		Ca					R	
5	Z						T	

(Period is labelled along the left vertical axis)

(4 *cont.***)**

By reference to this table, answer the following questions (use the letters given in the table when answering – you are not expected to identify the elements).

i Give the electronic arrangement of an atom of element G.

ii Which is the most reactive metal?

iii Which is the most reactive non-metal?

iv State, with a reason, which element J or L would react more vigorously with dilute hydrochloric acid. Write a balanced equation for the reaction between J and dilute hydrochloric acid.

v State, with a reason, which element X or L would react more vigorously with water. Write a balanced equation for the reaction between X and water.

vi Name the family of elements represented by M, Q, R and T.

vii If Q was bubbled through a solution of the potassium salt of T, state, with a reason, whether you would expect a reaction to occur.

viii Which element forms an amphoteric oxide? Given the formula of this oxide.

ix Elements X and E both form compounds with chlorine. Give the formula of each chloride and state the type of bonding found in each.

CHEMICAL FORMULAE AND BONDING

5 a Copy out and complete the table below.

b Explain, with the aid of diagrams, the electronic changes which occur when the following compounds are formed from their elements:

i Calcium nitride

ii Water.

(Atomic numbers: Ca = 20; N = 7; H = 1; O = 8)

Particle	Atomic number	Mass number	Number of			Electronic arrangement
			protons	neutrons	electrons	
Al atom	13			14		
Al^{3+} ion						
^{35}Cl isotope			17			
^{37}Cl isotope						
O^{2-} ion		16			10	
P atom		31				2.8.5

c State **two** physical properties you would expect **each** of the compounds in part **b** to exhibit. Account for each of these properties.

d Explain, with reference to the type of bonding present, why a typical metal is:

  i A solid at room temperature
  ii A good conductor of electricity
  iii Malleable and ductile.

THE STRUCTURE OF SOLIDS

6 a The table opposite gives some information about the physical properties of four solid substances. Suggest, with reasons, a possible structure for each of the substances, P, Q, R and S.

b Diamond and graphite are both composed of carbon atoms. How do you therefore account

(**6** *cont.*)

Substance	Melting point	Boiling point	Electrical conductivity	
			when solid	when molten
P	High	High	Good	Good
Q	High	High	Poor	Good
R	Low	Low	Poor	Poor
S	High	High	Poor	Poor

for the fact that diamond is a non-conductor and is extremely hard, whereas graphite is a good conductor and is soft and flaky?

SOLUTIONS, SOLUBILITY, SUSPENSIONS AND COLLOIDS

7 a Explain the difference between a solution, a suspension and a colloid, giving an example of each.

Temperature (°C)	12	26	40	56	65	78
Solubility of X (g per 100 g water)	19.5	36.5	60.0	95.5	121.0	173.5

Temperature (°C)	9	22	37	50	63	79
Solubility of Y (g per 100 g water)	79.5	88.5	102.0	114.5	129.0	157.0

b Using the data given in the tables above, draw the solubility curves for the two ionic solids X and Y. Both curves must be drawn on the **same** axes.

Use your graph to answer the following questions:

  i At what temperature is the solubility of the two solids the same?
  ii What mass of X would crystallise out if a saturated solution containing 200 g of water is cooled from 55°C to 35°C?
  iii What mass of Y must be added to a solution containing 1 dm^3 of water which

is saturated at 38°C in order to make a saturated solution at 48°C?

  iv If 8 g of X is heated with 10 g of water, at what temperature would a clear solution first be obtained?

c Using the following data, determine the solubility of substance A in water at room temperature:

           Mass of evaporating dish = 22 g

    Mass of evaporating dish +
           saturated solution of A = 99 g

    Mass of evaporating dish + solid A = 43 g

ACIDS, BASES AND SALTS

8 a Explain clearly:

   i the difference between an acid and a base

   ii the relationship between an alkali and a base.

b State **two** chemical properties, apart from their effect on indicators, which are typical of:

   i a dilute aqueous acid

   ii a dilute aqueous alkali.

   Illustrate **each** property by reference to a specific reaction — equations are essential.

c Hydrochloric acid and ethanoic acid are both acidic, however they have different pH values.

   i Describe how you would determine the pH value of each.

   ii Explain fully the reason for their different pH values.

d When a few drops of sodium hydroxide solution are added to lead(II) nitrate solution, a white precipitate of lead(II) hydroxide forms. On further addition of the sodium hydroxide solution, the precipitate disappears forming a colourless solution. Write a balanced equation for the formation of the precipitate and explain why it dissolves when excess sodium hydroxide solution is added.

9 a What is a salt?

b Distinguish between an acid salt and a normal salt, giving an example of each.

c Explain why some crystals, when heated, change their colour and shape even though they do not undergo any chemical change.

d Describe briefly, but including all essential experimental details and relevant equations, how you would prepare pure dry samples of the following:

   i Zinc nitrate starting from zinc carbonate

   ii Potassium sulphate starting from potassium hydroxide

   iii Barium sulphate starting from barium nitrate

   iv Anhydrous aluminium chloride starting from aluminium.

THE MOLE AND CHEMICAL CALCULATIONS

10 a i What is the mass of 0.3 mole of aluminium oxide?

   ii How many moles are there in 3.2 g of copper(II) sulphate?

   iii How many carbon dioxide molecules are there in 11 g of carbon dioxide?

   iv What mass of nitrogen contains the same number of molecules as 5.4 g of water?

b A compound with a relative molecular mass of 180 was found to contain 40% carbon, 6.7% hydrogen and 53.3% oxygen. Determine the molecular formula of this compound.

c What is the percentage by mass of oxygen in lead(II) nitrate?

11 a i How many moles are there in 224 cm^3 of oxygen at s.t.p.?

   ii What is the volume of 3.4 g of ammonia at r.t.p.?

   iii How many molecules of hydrogen are contained in 2.24 dm^3 of the gas at s.t.p.?

b Ethene burns in oxygen according to the equation:

$$C_2H_4(g) + 3O_2(g) \longrightarrow 2CO_2(g) + 2H_2O(g)$$

What volume of ethene, measured at r.t.p., would burn completely in 10.8 dm^3 of oxygen?

c If a solution of sodium carbonate containing 12.0 g of carbonate ions reacts completely with hydrochloric acid, what volume of carbon dioxide would be produced at s.t.p.?

12 a What mass of potassium carbonate is needed to prepare 250 cm^3 of a solution of concentration 0.2 mol dm^{-3}?

b How many moles of sodium hydroxide are present in 400 cm^3 of a solution which has a concentration of 2 g dm^{-3}?

c Copper(II) nitrate can be prepared by adding copper(II) oxide to nitric acid. What is the maximum mass of copper(II) nitrate which can be formed from 150 cm^3 of nitric acid of concentration 0.5 mol dm^{-3}?

13 a When copper(II) oxide is heated with carbon, copper and carbon dioxide are produced. What mass of copper(II) oxide is required to react completely with 4 g of carbon?

b What mass of potassium hydrogen carbonate must be added to sulphuric acid to produce 8.7 g of potassium sulphate?

c When lead(II) nitrate is heated the following reaction occurs:

$$2Pb(NO_3)_2(s) \longrightarrow 2PbO(s) + 4NO_2(g) + O_2(g)$$

If 3.31 g of lead(II) nitrate are heated in a test tube until no further reaction takes place, calculate the decrease in mass of the contents of the tube.

QUANTITATIVE ANALYSIS CALCULATIONS

14 a 25 cm³ of sulphuric acid of concentration 0.2 mol dm⁻³ are needed to exactly neutralise 40 cm³ of sodium hydroxide solution. What is the concentration of the sodium hydroxide solution in mol dm⁻³?

b It was found that 7.5 cm³ of hydrochloric acid of concentration 2.0 mol dm⁻³ are required to neutralise 15 cm³ of ammonium carbonate solution of concentration 48 g dm⁻³. Show fully how this information can be used to write an equation for the reaction.

c 50 cm³ of an acid with the formula H_nX and concentration 0.4 mol dm⁻³ are needed to neutralise 100 cm³ of potassium carbonate solution of concentration 13.8 g dm⁻³. Calculate the value of n in H_nX.

TYPES OF CHEMICAL REACTIONS

15 a State, with a reason in **each** case, what type of reaction you consider each of the following to be. Some may be of more than one type.

i $2Pb(NO_3)_2(s) \longrightarrow$
$$2PbO(s) + 4NO_2(g) + O_2(g)$$

ii $ZnO(s) + 2HCl(aq) \longrightarrow$
$$ZnCl_2(aq) + H_2O(l)$$

iii $2Mg(s) + O_2(g) \longrightarrow 2MgO(s)$

iv $BaCl_2(aq) + H_2SO_4(aq) \longrightarrow$
$$BaSO_4(s) + 2HCl(aq)$$

v $N_2(g) + 3H_2(g) \rightleftharpoons 2NH_3(g)$

b The following represent displacement reactions which may or may not actually take place:

i $2Zn(s) + Pb(NO_3)_2(aq) \longrightarrow$
$$2ZnNO_3(aq) + Pb(s)$$

ii $Cl_2(g) + 2KBr(aq) \longrightarrow$
$$2KCl(aq) + Br_2(aq)$$

iii $3Cu(s) + 2AlCl_3(aq) \longrightarrow$
$$3CuCl_2(aq) + 2Al(s)$$

For each reaction state, with a reason, whether or not the reaction would occur.

OXIDATION AND REDUCTION

16 a Define the following in terms of oxidation number:

i An oxidising agent
ii A reducing agent.

b State, with reasons, whether sulphur dioxide is acting as an oxidising agent or a reducing agent in each of the following reactions:

i $2H_2S(g) + SO_2(g) \longrightarrow 2H_2O(l) + 3S(s)$

ii $SO_2(g) + H_2O(l) + NaClO(aq) \longrightarrow$
$$NaCl(aq) + H_2SO_4(aq)$$

c Oxidation is often defined as an increase in the oxidation number of an element due to the loss of electrons, and reduction as the reverse of oxidation. Explain how this statement applies in each of the following reactions:

i The conversion of iodide ions to iodine
ii The conversion of iron(III) ions to iron(II) ions.

Equations are essential.

d You are provided with two solutions, A and B. A is thought to be an oxidising agent and B a reducing agent. Describe **two** tests that you could carry out to confirm the above suspicions.

ELECTROCHEMISTRY AND ELECTROLYSIS

17 a Distinguish between metallic conduction and electrolytic conduction.

b Explain why a solution of sodium hydroxide, having the same concentration as a solution of ammonia, is a better conductor.

c **i** Explain clearly why, when copper(II) sulphate solution is electrolysed using carbon electrodes, oxygen is produced at the anode but a different result is

153

obtained when a copper anode is used. Your answer must include relevant equations.

ii If a current of 1.5 amperes flows for 48 minutes 15 seconds through a solution of copper(II) sulphate, what mass of copper would be deposited at the cathode?

d

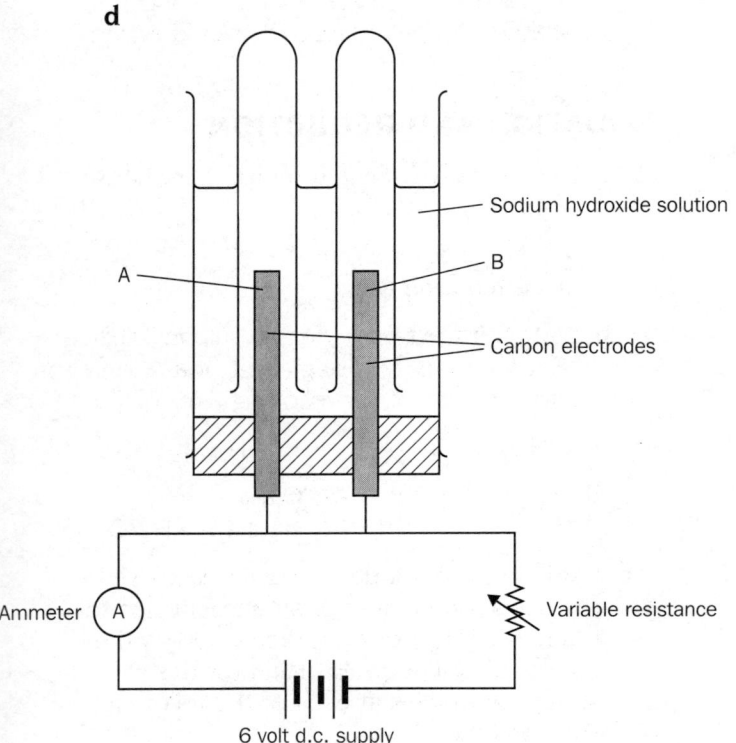

6 volt d.c. supply

A simple electrolysis cell was set up as shown above.

i   Identify the electrodes A and B.
ii What would you expect to happen at each electrode? Your answer must include relevant equations and a consideration of the relative volumes of gases produced.
iii What change, if any, would you expect in the electrolyte? Give a reason for your answer.
iv How long must a steady current of 0.5 amperes flow through the circuit in order to produce 11.2 cm^3 of gas at the anode, the volume being measured at s.t.p.?

ENERGY AND CHEMICAL ENERGETICS

18 a When we burn coal, we are making use of energy which came originally from the Sun. Explain fully the reason for this.

b Man today is constantly trying to develop alternative energy sources.

i   Give **two** reasons for this.
ii Name **two** alternative energy sources which you feel would be suitable for use in the Caribbean. Describe how **each** of these may be used. Your answer should include any disadvantages associated with the use of each energy source described.

c When magnesium reacts with hydrochloric acid, there is a noticeable increase in temperature.

i   Is the reaction endothermic or exothermic?
ii Account for the temperature change in terms of enthalpy of the reactants and products.
iii Draw an energy profile diagram for the reaction.

d Account for the fact that the heat of neutralisation of sodium hydroxide by nitric acid is the same as the heat of neutralisation of potassium hydroxide by sulphuric acid.

e Burning 480 cm^3 of propane (C_3H_8), measured at r.t.p., caused the temperature of 250 cm^3 of water to increase by 42.5 °C. Calculate the heat of combustion of propane. (1 mole of any gas at r.t.p. occupies 24 dm^3; specific heat capacity of water = 4.18 J g^{-1} °C^{-1}.)

RATES OF REACTION

19 The graph below shows the results of an experiment to determine the rate of reaction between magnesium ribbon and dilute hydrochloric acid.

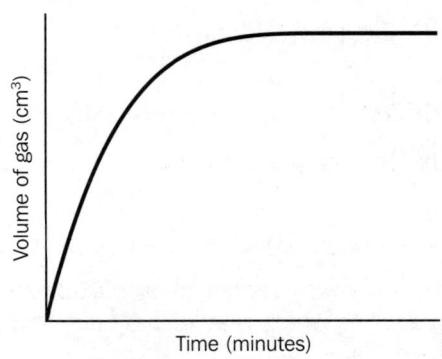

a Write an equation for the reaction and identify the gas produced.

b Account for the shape of the curve.

c If 4 g of magnesium ribbon and 50 cm^3 of hydrochloric acid of concentration 0.5 mol dm^{-3} are used, when the reaction stops what effect would you expect the contents of the flask to have on:

  i Red litmus paper
  ii Blue litmus paper?

Explain your answers fully.
(Relative atomic masses: H = 1; Mg = 24; Cl = 35.5)

d What effect, if any, would using 4 g of magnesium filings instead of ribbon have on:

  i The rate of evolution of the gas
  ii The total volume of gas evolved?

Explain your answers fully.

e What effect, if any, would using 100 cm^3 of hydrochloric acid of concentration 0.5 mol dm^{-3} have on:

  i The rate of evolution of the gas
  ii The total volume of gas evolved?

Explain your answers fully.

METALS AND THEIR COMPOUNDS

20 a Describe **three** different chemical properties of metals, using a different metal to illustrate each property (equations are essential).

b Experiments carried out with three metals, X, Y and Z, gave the following results:

- X displaced copper from an aqueous solution of copper(II) nitrate, but did not react with hydrochloric or sulphuric acid.
- Y reacted vigorously with cold water.
- Z reduced iron(III) oxide but not aluminium oxide.

  i Based on these results and your knowledge of the reactivity series of metals, place the metals X, Y, Z, copper, iron, aluminium and magnesium in **decreasing** order of reactivity. (You are not required to identify X, Y and Z.)
  ii If X, Y and Z are all divalent, write equations for the reactions between X and copper(II) nitrate solution, Y and cold water, and Z and iron(III) oxide.

(Use X, Y and Z as the symbols for the metals.)

  iii For each of the metals Y and Z state, with reasons, the method you consider would be the most suitable for extracting the metal from its ore.

c What effect, if any, would heat have on the following compounds:

  i Copper(II) carbonate
  ii Lead(II) nitrate
  iii Sodium hydroxide?

Write balanced equations where appropriate.

21 a Aluminium and iron are both extracted on a large scale from their oxides. Name these oxides and give an account of **both** extraction processes, clearly emphasising the chemical principles involved and giving equations where appropriate.

b Give reasons for each of the following:

  i Aluminium is often used in cooking utensils.
  ii Zinc is used as a protective coating for iron.
  iii Duralumin is used in preference to aluminium in the construction of aircraft bodies.

c 'Metals are essential to the life of living organisms. They can, however, be extremely harmful to the same organisms.' Discuss this statement by reference to specific metals.

NON-METALS AND THEIR COMPOUNDS

22 a 'Non-metals are usually considered to be oxidising agents, however some may also behave as reducing agents.' Support this statement by reference to the reactions of hydrogen, carbon, oxygen and chlorine with metals and/or metal oxides (equations are essential).

b The following observations apply to three different oxides, A, B and C:

- A is a white solid which is insoluble in water, but forms colourless solutions when mixed with both hydrochloric acid and sodium hydroxide solution.
- B is a colourless gas which reacts with water to form a solution which turns blue litmus to red.

— C is a white solid which reacts with water to form a solution which turns red litmus to blue.

   i With reasons, classify each oxide.

   ii In each case name **one** oxide which has the properties indicated.

c 'Non-metals and their compounds are used extensively by Man.' By reference to **three** different non-metals, provide evidence to support this statement.

23 a Discuss the importance of recycling both in nature and by Man in today's world. Support your answer by reference to specific examples.

b Illustrate, by means of a simple diagram, how water is recycled in nature.

c Explain in detail how a nitrogen atom can begin in the air and eventually return to the air having passed through a hibiscus plant and a tortoise.

d Fossil fuels are essential to life as we know it. Discuss some of the possible consequences if Man continues his indiscriminate use of fossil fuels.

LABORATORY PREPARATION AND IDENTIFICATION OF GASES

24 a i Draw a labelled diagram of the apparatus you would use to prepare and collect dry carbon dioxide in the laboratory.

   ii Name the reagents and drying agent, and state the conditions you would use for the above preparation.

   iii Give an equation for the reaction.

b A solution of chlorine in water may be used to bleach clothes. However, if the solution is too concentrated, holes often appear in the fabric. Explain **each** of these facts.

c On bubbling an unknown gas X through limewater, a white precipitate formed. Further bubbling caused the limewater to become colourless once more. Suggest an identity for gas X and account for the observations given. Your answer must include relevant equations.

d On bubbling another unknown gas Y through acidified potassium manganate(VII) solution, the latter turned colourless. Suggest an identity for gas Y and explain

the chemical principles involved in the colour change.

SOME INDUSTRIAL PROCESSES

25 a Describe how ammonia is manufactured from natural gas. Your answer must include the chemical principles, conditions and equations for the manufacturing process.

b The manufacture of sulphuric acid involves the production of sulphur trioxide.

   i Describe, giving relevant equations, how sulphur trioxide is produced.

   ii In converting sulphur trioxide to sulphuric acid, the sulphur trioxide cannot be added directly to water. Explain why.

c i Describe how chlorine and sodium hydroxide are manufactured from brine. Your answer must include all relevant equations.

   ii State, with reasons, the nature of each electrode used in the manufacturing process described above.

QUALITATIVE ANALYSIS

26 a For each of the following pairs of substances, describe a chemical test you could use to distinguish between them. Your answers must include the results of the test on both substances in each pair, and equations where relevant.

   i Calcium nitrate and zinc nitrate

   ii Potassium chloride and potassium iodide

   iii Aluminium chloride and ammonium chloride

   iv Sodium sulphate and sodium sulphite.

b When a white powder, A, was heated in a test tube a gas, B, which formed a milky white precipitate in limewater, was evolved and a yellow residue, C, remained in the tube. When dilute nitric acid was added to the tube, C dissolved without effervescence to form a colourless solution, D. On adding a few drops of dilute sodium hydroxide solution to a portion of D, a white precipitate, E, formed which dissolved in excess sodium hydroxide solution. On adding a few drops of dilute hydrochloric acid to a second portion of D, a white precipitate, F, formed which disappeared when the tube was heated.

    i With reasons and relevant equations, identify each of the substances A, B, C, D, E and F.

    ii Describe a confirmatory test you could carry out for the anion of substance A.

c When a sample of a solid, X, was heated in a test tube, a brown gas was given off and a glowing splint inserted into the tube relit. A second sample of X was made into a solution by adding distilled water. On adding dilute aqueous ammonia to this solution, a blue precipitate formed. On further addition of the dilute ammonia solution, the precipitate disappeared forming a deep blue solution. Explaining these observations and giving relevant equations, identify X.

ORGANIC CHEMISTRY

27 a Explain, with the aid of a diagram, the bonding between the atoms in a molecule of ethane, C_2H_6.

   b Two non-cyclic hydrocarbons Y and Z have empirical formulae CH_2 and C_3H_7 respectively, and relative molecular masses of 56 and 86 respectively.

    i To which homologous series does Y belong? Give a reason for your answer.

    ii Name both Y and Z, and give the molecular formula of each.

    iii Write down the full structural formulae of **two** isomers of Y and **three** isomers of Z.

    iv Describe a chemical test which you could use to distinguish between Y and Z. Your answer must include the results of the test on both Y and Z.

   c Ethane is said to be a 'saturated' hydrocarbon and ethene an 'unsaturated' hydrocarbon.

    i How do the structures of ethane and ethene affect their reactivity? Illustrate your answer by comparing the reactivity of ethane and ethene with chlorine in diffused light.

    ii Write an equation for the reaction of ethene and chlorine.

    iii Describe **two** other addition reactions of ethene, excluding those with other halogens. Your answer must include the

conditions and an appropriate equation for each reaction.

   d Name **four** fractions obtained by the fractional distillation of crude oil and give **one** major use of each fraction named.

28 a Describe how ethanol may be prepared in the laboratory from a carbohydrate. Include an equation for the reaction and a description of the method of purification of the ethanol.

   b Describe how each of the following may be formed from ethanol:

    i Ethene
    ii Ethyl ethanoate.

   Include in your answers the reaction conditions and relevant equations.

   c A housewife found that her wine had gone sour. Explain, with an appropriate equation, how this occurred.

   d The breathalyser test is used to detect drunken drivers.

    i Explain the chemical principles behind this test.

    ii One of your close relatives is an alcoholic. Give **three** reasons you could use to try to stop him drinking.

29 a 'An aqueous solution of ethanoic acid reacts in a similar way to the common mineral acids found in the laboratory.' Support this statement by reference to **two** different chemical reactions of dilute aqueous ethanoic acid. Your answer must include appropriate equations.

   b A compound, X, with the formula

$$CH_3CH_2CH_2-\overset{\displaystyle O}{\overset{\displaystyle \|}{C}}-O-CH_3$$

is heated with dilute hydrochloric acid.

    i Name the group of organic compounds to which X belongs.

    ii Give the names and formulae of the products of the above reaction.

    iii Name the type of reaction involved.

   c One use of vegetable oils and animal fats is in the manufacture of soap.

    i How is soap produced from the fat glyceryl tristearate? Give an equation for the reaction involved and name the type of reaction.

ii Explain how soap acts to remove grease from clothing.

iii Today, many soapless detergents are available. If you had the choice of using a soapy or a soapless detergent, which, would you choose and why?

POLYMERS

30 a What do you understand by the term 'polymerisation'?

b Draw the structure of a polymer which could be obtained from the following:

i

$$HOOC-X-COOH \quad \text{and} \quad HO-Y-OH$$

ii

In **each** case name the type of polymer formed, the type of linkage it contains and give **two** uses.

c Polysaccharides are naturally occurring polymers.

i Show how a polysaccharide molecule is built up from its monomer units.

ii Describe **two** ways by which polysaccharide molecules can be converted to their monomer units.

d By reference to a **named** polymer, show how its physical and chemical properties are different to those of the monomer units from which it is formed.

e Synthetic polymers, often referred to as 'plastics', are used increasingly in today's world. Indicate some of the advantages and disadvantages of this trend.

Relative atomic masses

These elements are referred to in calculations.

$H = 1$	$Na = 23$	$Cl = 35.5$
$C = 12$	$Mg = 24$	$K = 39$
$N = 14$	$Al = 27$	$Cu = 64$
$O = 16$	$S = 32$	$Pb = 207$

NUMERICAL ANSWERS TO REVISION QUESTIONS

7 b i 70 °C
  ii 83 g
  iii 95 g
  iv 50 °C

  c 37.5 g per 100 g water

10 a i 30.6 g
  ii 0.02 mole
  iii 1.505×10^{23} molecules
  iv 8.4 g

  b $C_6H_{12}O_6$
  c 29.003%

11 a i 0.01 mole
  ii 4.8 dm^3
  iii 6.02×10^{22} molecules

  b 3.6 dm^3
  c 4.48 dm^3

12 a 6.9 g
  b 0.02 mole
  c 7.05 g

13 a 53.333 g
  b 10 g
  c 1.08 g

14 a 0.25 mol dm^{-3}
  b 0.015 mole HCl neutralises 0.0075 mole $(NH_4)_2CO_3$
  2 moles HCl neutralise 1 mole $(NH_4)_2CO_3$
  i.e. equation based on the above:

  $$(NH_4)_2CO_3(aq) + 2HCl(aq) \longrightarrow$$
  $$2NH_4Cl(aq) + H_2O(l) + CO_2(g)$$

  c $n = 1$

17 c ii 1.44 g
  d iv 386 seconds or 6 minutes 26 seconds

18 e 2221 kJ mol^{-1}

19 c Number of moles of reactants present:
  0.025 mole HCl and 0.167 mole Mg
  0.0125 mole Mg reacts completely with 0.025 mole HCl

  $\therefore$ All HCl is used up, Mg is present in excess.

Index

acid anhydrides   39
acids   37–9
    basicity of   38
    properties of   37–8
    recognition of   40
    strength of   40–1
alcohol abuse   138
alcohols   136–8
alkalis   39–40
    properties of   40
    recognition of   40
    strength of   40–1
alkanes   131–2, 135
alkanoic acids   138–40
alkenes   132–5
allotropes
    of carbon   28–9
    of sulphur   104
alloys   98
aluminium
    corrosion of   97
    extraction of   94–5
    uses of   97
ammonia
    laboratory preparation of   113
    manufacture of   117
    uses of   102
amphoteric substances   41–2
anodising   74
antacids   39
atoms   4, 12–15
    mass of   15
    structure of   12–14
Avogadro constant   47
Avogadro's Law   49

baking powder   39
bases   39
    solubility of   43
bleach   105, 116
bonding   20–7
Brownian motion   3

carbon
    allotropes of   28–9
    cycle   108

dating   14
    properties of   101
    uses of   101
carbon dioxide
    laboratory preparation of   112–13
    uses of   39, 101
changing state   2
chemical reactions, types of   60–1
chlorine
    manufacture of   115
    properties of   105
    uses of   106
chromatography   8
colloids   36
compounds   6
Contact process   116–17
corrosion of metals   97
covalent
    bonding   24–6
    compounds, properties of   27
cracking   129
crystallisation   7

detergents   142–3
diamond
    properties of   29
    structure of   28
    uses of   101
diffusion   3
displacement reactions   60, 90
distillation
    fractional   10
    of crude oil   129–30
    simple   9

electrochemical series   93
electrolysis   68–77
    of aqueous solutions   70–2
    of molten substances   70
    quantitative   75–7
    uses of   73–5
electroplating   74
elements   5–6
endothermic reactions   80–2
energy   78–84
    activation   82, 85, 87

energy (*continued*)
    alternative sources of   79–80
    in chemical reactions   80–4
    interconversions   78
    major sources of   78–9
enthalpy   81
equations
    how to write   31–3
    ionic   32–3
esters   140–2
ethanoic acid   139–40
ethanol   136–8
ethene   134
ethyl ethanoate   140, 141
evaporation   7
exothermic reactions   80–82

Faraday's Laws of Electrolysis   75, 76
filtration   7
formulae
    how to determine quantitatively   51–2
    how to write   20–1, 23–4
    of organic compounds   126
fuels   78–9

gases
    identification of   114
    laboratory preparation of   112–13
    properties of   1
graphite
    properties of   29
    structure of   29
    uses of   101

Haber process   117
heat of
    combustion   83–4
    neutralisation   82–3
    solution   84
homologous series   127–8
hydrocarbons   129–35
hydrogen   100–1

indicators   40, 41
ionic
    bonding   21–2, 26
    compounds, properties of   27
ions   4, 21–4
    tests for   119–24
iron
    corrosion of   97
    extraction of   95–6
    uses of   97
isomerism   126–7
isotopes   13–15

Law of
    Conservation of Energy   78
    Conservation of Matter   53
    Constant Composition   51
lead
    pollution by   99
    uses of   98
liquids, properties of   1

matter   1–4
    particulate theory of   2–4
    states of   1–2
metal compounds
    effect of heat on   92
metals
    bonding in   27
    extraction of   94–6
    pollution by   99
    properties of   6, 27, 89
    reactions of   90–1
    reactivity series of   89–90
    uses in living organisms   98–9
    uses of   97–8
mixtures   6
    separating   7–10
mole   47
    and chemical formulae   51–3
    and chemical reactions   53–5
    and gas volumes   49–50
    and mass   47–49
    and quantitative analysis   56–9
    and quantitative electrolysis   75–7
    and solutions   50–1
molecules   4

nitric acid
    manufacture of   118
    uses of   102
nitrogen
    cycle   108–9
    manufacture of   116
    properties of   102
    uses of   102
non-metal compounds
    pollution by   110–11
    uses of   101–2, 105, 106
non-metals
    properties of   6, 100–5
    reactions of   100–5
    uses in living organisms   109–10
    uses of   101, 102, 105, 106

organic compounds
    formulae of   126
    structure of   125–6
osmosis   4

oxidation   62–7
    number   62–3
    recognition of   64–5
oxides, classification of   103–4
oxidising agents   65–7
    tests for   67
oxygen
    laboratory preparation of   112–13
    manufacture of   116
    properties of   103

Periodic Table   x, 16–19
    trends in Group II   18
    trends in Group VII   18
    trends in Period 3   19
pH scale   41
phosphorus, uses of   106
pollution   99, 110–11
polymers   144–8
polysaccharides   144, 146, 147
proteins   144, 145, 147
pure substances   6

qualitative analysis   119–24
quantitative analysis   56–9

radioisotopes   14–15
rate curves   88
rates of reaction   85–8
    factors affecting   85–7
    measurement of   85
reducing agents   65–7
    tests for   67
reduction   62–7
    recognition of   64–5
relative atomic mass   15, 46
relative formula mass   46, 47
relative molecular mass   46–47
reversible reactions   60–1

salts   42–5
    preparation of   43–5
    solubility of   43
    types of   42
separating funnel   8
silicon, uses of   106
soap
    action of   142–3
    preparation of   142
sodium chloride
    bonding in   22
    properties of   29
sodium hydroxide
    manufacture of   115
solids
    properties of   1
    structure of   28–30
solubility   34–6
solutions   34
solvent extraction   8
sublimation   2, 9
sucrose   148
    extraction of   10–11
sulphur
    allotropes of   104
    properties of   104
    uses of   105
sulphuric acid
    manufacture of   116–17
    uses of   105
suspensions   36

valency   20–21

water
    cycle   107
    of crystallisation   43